Mit „BestMasters" zeichnet Springer die besten Masterarbeiten aus, die an renommierten Hochschulen in Deutschland, Österreich und der Schweiz entstanden sind. Die mit Höchstnote ausgezeichneten Arbeiten wurden durch Gutachter zur Veröffentlichung empfohlen und behandeln aktuelle Themen aus unterschiedlichen Fachgebieten der Naturwissenschaften, Psychologie, Technik und Wirtschaftswissenschaften. Die Reihe wendet sich an Praktiker und Wissenschaftler gleichermaßen und soll insbesondere auch Nachwuchswissenschaftlern Orientierung geben.

Springer awards "BestMasters" to the best master's theses which have been completed at renowned Universities in Germany, Austria, and Switzerland. The studies received highest marks and were recommended for publication by supervisors. They address current issues from various fields of research in natural sciences, psychology, technology, and economics. The series addresses practitioners as well as scientists and, in particular, offers guidance for early stage researchers.

More information about this series at https://link.springer.com/bookseries/13198

Sarah Marie Treibert

Mathematical Modelling and Nonstandard Schemes for the Corona Virus Pandemic

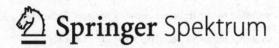

Springer Spektrum

Sarah Marie Treibert
Wuppertal, Germany

ISSN 2625-3577   ISSN 2625-3615 (electronic)
BestMasters
ISBN 978-3-658-35931-7   ISBN 978-3-658-35932-4   (eBook)
https://doi.org/10.1007/978-3-658-35932-4

Responsible Editor: Marija Kojic
This Springer Spektrum imprint is published by the registered company Springer Fachmedien Wiesbaden GmbH part of Springer Nature.
The registered company address is: Abraham-Lincoln-Str. 46, 65189 Wiesbaden, Germany

Contents

Abbreviations

ACE2	Angiotensin-converting enzyme 2
ARDS	Acute respiratory syndrome
BARDA	Biomedical Advanced Research and Development Authority
CFR	Case-fatality rate
COPD	Chronic obstructive pulmonary disease
COVAX	COVID-19 Vaccines Global Access organization
COVID-19	Corona virus disease 2019
DDE	Delay Differential Equation
DFE	Disease-free equilibrium
DNA	Desoxyribonuclein acid
ECDC	European Centre for Disease Prevention and Control
EE	Endemic equilibrium
ELISA	Enzyme-linked immunosorbed essay
GN	Gauss-Newton Method
IC	Initial conditions
ICOSARI	ICD-10-code-based hospital sentinel for severe acute respiratory diseases
ICU	Intensive care unit
IFR	Infection-fatality rate
IgG	Immunglobulin G
ILI	Influenza-like illness
MCMC	Markov Chain Monte Carlo Method
MERS-CoV	Middle mast respiratory syndrome corona virus
mRNA	messenger ribonucleic acid
NGM	Next-generation matrix
NLS	Nonlinear least squares approach

NPI	Non-pharmaceutical intervention
NSFD	Nonstandard finite difference scheme
OxCGRT	Oxford COVID-19 Government Response Tracker
PDE	Partial Differential Equation
PDF	Probability density function
PHA	Public Health Agency of Sweden
RBD	receptor-binding domain
RKI	Robert-Koch Institute
RNA	Ribonuclein acid
RT-PCR	Reverse transcriptase polymerase chain reaction
SARI	Severe acute respiratory disease
SARS-CoV̇	Severe acute respiratory syndrome corona virus type
SARS-CoV-2	Severe acute respiratory syndrome corona virus type 2
STIKO	German Standing Committee on Immunisation
System of ODEs	System of ordinary differential equations
WHO	World Health Organization
2019-nCoV	Novel corona virus, discovered in the year 2019

List of Figures

List of Tables

Introduction

1

This introduction is divided into the positioning of this thesis in the thematic background of the novel corona virus in Section 1.1 and more exact explanations of the contributions of this thesis in Section 2.1.

1.1 Thematic Background

The first cases of the previously unknown severe acute respiratory syndrome coronavirus 2 (SARS-CoV-2) appeared in China at the end of the year 2019, but were not recognized as infections with the novel corona virus (2019-nCov) back then. The genetic sequence of SARS-CoV-2 was identified in early January 2020. Many researchers have shown that SARS-CoV-2 has a zoonotic source [1]. It has not yet been disproved that the virus originated from a laboratory accident, although this thesis is improbable [2, 3]. Environmental samples taken from the Huanan Wholesale Seafood Market in Wuhan City, where seafood, wild, and farmed animal species were sold in December 2019, were tested positive for SARS-CoV-2, implying that this market was the source or played a role in the initial amplification of the outbreak [1]. The disease effected by the SARS-CoV-2-infection is called the corona virus disease 2019 (COVID-19). Originating from China, the virus spread across the whole world via air-borne virus infection from the end of January 2020.

Between December 31st 2019 and the 12th calendar week in 2021, $127, 628, 928$ cases and $2, 791, 055$ deaths were registered worldwide [4]. The first case in the United States was reported on January 22nd 2020, in France on January 24th, in Germany on January 27th, in Italy and the United Kingdom on January 31st, and in Sweden and Spain on February 1st 2020 [5]. South America was reached in February 2020 in order that the later strongly affected countries Brazil, Mexico, Colombia reported their first SARS-CoV-2 case on February 26th, February 28th or March

© The Author(s), under exclusive license to Springer Fachmedien Wiesbaden GmbH, part of Springer Nature 2021
S. M. Treibert, *Mathematical Modelling and Nonstandard Schemes for the Corona Virus Pandemic*, BestMasters, https://doi.org/10.1007/978-3-658-35932-4_1

6th, respectively [5]. Governments around the globe took measures to combat the infectious disease in their own countries, which included temporary lockdowns of the population and shut-downs of certain production activities [6]. The rapid spread of the virus put the intensive and acute care stations in hospitals in Western Europe and the United States to the test. Intensive care capacities were fully exhausted in some regions of Italy, France, Spain and the United Kingdom between April and early May 2020 [6].

According to the European Centre for Disease Prevention and Control (ECDC), 3.32% of all confirmed cases were reported from Africa, 18.84% from Asia, 44.21% from America, 33.57% from Europe, 0.05% from Oceania and 0.00057% from an international conveyance in Japan as of the calendar week 11 in 2021 [4]. As of March 31st 2021, the countries with the highest total numbers of confirmed infections in the world were the United States (31, 097, 154), Brazil (12, 664, 058), India (12, 149, 335), France (4, 585, 385), Russia (4, 536, 820) and the United Kingdom (4, 341, 736) [7]. The most affected European countries apart from France and the United Kingdom at that time were Italy with 3, 561, 012, Spain with 3, 275, 819 and Germany with 2, 809, 510 confirmed infections [7]. As of March 21st 2021, the ten European countries with the largest numbers of infections per 100, 000 inhabitants in descending order of incidence height were Montenegro, Czechia, Slovenia, Luxembourg, Portugal, Serbia, Lithunia, Sweden, Belgium and Estonia [8].

The development of COVID-19 vaccines has been promoted by governments collaborating with pharmaceutical companies since early 2020. The United Kingdom was the first country worldwide to approve the COVID-19 vaccine of the company BioNTech/Pfizer on December 2nd 2020, followed by the United States on December 11th and the European Union on December 21st 2020 [9]. The vaccines of the US-American company Moderna, the Swedish-British company AstraZeneca, and the Belgian company Janssen were authorized as the second, third or fourth COVID-19 vaccine in the European Union on January 6th, January 21st or March 11th 2021, respectively. Multiple other COVID-19 vaccines are currently developed, and a considerable part is examined in phase III studies [9]. The development of effective vaccines, the efficient distribution of vaccines and the fast and safe vaccination of large parts of populations worldwide is a current worldwide challenge. In the second half of the year 2020, the first remarkable SARS-CoV-2 mutations were detected in the United Kingdom, followed by a variant spreading in South Africa and one found in Brazil in December 2020. In particular, the virus lineage B.1.1.7, that widely spread across the United Kingdom in December 2020, is a variant of great concern in Europe due to its large number of mutations [10]. In March 2021, another variant including two different mutations was detected in India. SARS-CoV-2 mutations represent a major challenge in the worldwide combat against the pandemic due to a presumed and partly proved easier viral transmissibility and a connected higher infectiousness and mortality risk.

A large number of scientific modelling approaches covering different fields of expertise and pursuing distinct targets with respect to predicting and controlling the progression of the pandemic have been published since the outbreak of 2019-nCoV. A comparatively often discussed topic is optimal authoritarian intervention measures that should reduce infection numbers to prevent a future exponential growth, but not restrict economies and the lives of population members too invasively.

A challenge in infectious disease modelling is finding and implementing an appropriate mathematical model, which includes certain transmission and transition dynamics observable in the real world. Suitable and preferably exact algorithms and estimation methods to fit the model to available data like infection, hospitalization and death numbers and make forecasts must be selected and applied.

With regard to these two aspects, a deterministic model fundamentally based on the well-known SIR model is developed in this thesis in order to make compartment size predictions by model parameter modifications reflecting distinct degrees of intervention measures like contact restrictions, non-pharmaceutical interventions (NPIs), quarantine, isolation and vaccination programs. The model development is substantiated with the current level of information concerning the incidence, symptomatology, disease progression, governmental measures and lethality of SARS-CoV-2 and COVID-19 and some enhanced theory of compartment models. These two fundamentals are combined to make extended deliberations with respect to the compartment structure and transmission and transition dynamics necessary to model the spread of SARS-CoV-2.

1.2 Methods, Structure and Objectives

This section firstly gives a description of the main contributions used in this thesis in Subsection 1.2.1. Therefore, the methods of this work are characterized, and it is described what are no aims of this work. In Subsection 1.2.2, the structure of this thesis is elucidated. Subsection 1.2.3 summarizes the main targets of this work.

1.2.1 Methods

With respect to mathematical derivations, the theory behind systems of ordinary differential equations (systems of ODEs), initial or boundary value problems and their solutions are not a part of this thesis. Beside this, different methods of analytically solving or numerically integrating systems of ODEs are not analytically compared. Conversely, the **largest contributions** of this thesis are the following two:

The first one is the development of a complex compartment model, through which specific scenarios of transitions between population groups in times of the COVID-19 pandemic can be optimally modelled.

The complex model that variants are derived from is sometimes denoted as the enhanced or core model throughout this thesis. It is meant to serve as a foundation for the adaptation to specific COVID-19 modelling goals in subsequent parts of this thesis as well as further scientific approaches. The assumptions of a fundamental compartment model named named *SIR model* and compartment enhancements often used in epidemiology are explained in Chapter 3. Building on this, the assumptions, compartments and definitions of inter-compartmental dynamics in the enhanced model are presented in Chapters 4 and 5. It has to be explained at this point that a *compartment* represents a population group distinguished by a certain stage of progression of the observed infection, which can for instance include infectiousness, symptom development, treatment and the absence of infectedness. The second large contribution of this thesis is the implementation of two model variants in $MATLAB$ for the purpose of evaluating the influence of certain model parameters on model predictions concerning future progressions of the COVID-19 spread through parameter calibration.

Thematically, this thesis does not analyse political decisions which have been made or societal or economic questions that have risen during the corona virus pandemic. However, these issues are partly, briefly addressed in Chapter 2.

1.2.2   Structure

This thesis has the following **structure**:

In **Chapter** 2 in Section 2.1, worldwide SARS-CoV-2 incidences are shown and assessed with the aid of publicly available graphs in order to provide a foundation for the evaluation of the numbers of new infections. The aetiology and transmission paths of SARS-CoV-2, symptomatology and disease progression of COVID-19 are explained on a rather biological level in Sections 2.2 and 2.3. Aside from that, realized policies of containing the spread of SARS-CoV-2 and COVID-19 including testing, vaccinations and non-pharmaceutical interventions (NPIs) are presented in Section 2.4, in particular with regard to Germany and Sweden. Moreover, case-fatality rates and possible errors in their computation are explained in Section 2.5 for the purpose of understanding the magnitudes of the case-fatality rates used in the later implementations.

The theory of systems of ODEs is explained and used in **Chapter 3** to introduce the theory of compartment models in epidemiology, and derive different systems of ODEs corresponding to the core compartment model or one of its variants in Chapter 5. During the process of establishing a complex compartment model that optimally fits the population groups shown in times of the COVID-19 pandemic, multiple scientific papers that also deal with modelling this pandemic have been seen through and examined.

It has to be stressed that the compartment model of Chapters 4 and 5 is independently deduced and established based on certain considerations. These considerations concern the division of populations into groups characterized by distinct degrees of affectedness, that have become visible in various countries during the pandemic, and the transitions of individuals from one to another compartment, that are both characterized by distinct features that the individuals in them exhibit. For instance, such a feature can be infectivity, showing symptoms, being hospitalized or having recovered.

Chapter 4 is structured into three sections, that deal with the differences between a basic compartment model presented in Chapter 3 and the enhanced compartment model, a detailed description and overview of the compartments and inter-compartmental transitions, and the derivation and definition of the transmission rates used in the model implementations.

In **Chapter 5**, compartment modifications are derived from the core compartment model, and the systems of ODEs corresponding to two model variants and the core model are established. Moreover, different kinds of so-called *reproduction numbers* are defined.

Chapter 6 deals with the $MATLAB$ implementations of two compartment model variants. The implementations are prepared with the aid of the formulation of the parameter optimization method that is realized in $MATLAB$, a description of the whole implementation process, and an overview of the definitions of all parameters and data sources per model variant.

The main target of the implementation of each model is the minimization of the error between compartment size data of successive points in time, which are obtained from reliable publicly available sources, and compartment size data obtained from the integration of a used system of ODEs from a determined initial to a final point in time. This error minimization is accompanied by the optimization of model parameters like contact, quarantine and transmission rate parameters, to adapt a certain compartment model to distinct scenarios of state interventions, optionally within parameters bounds. It can mathematically be realized with the aid of different optimization algorithms. The method of *nonlinear least squares* (NLS) is derived and applied in the $MATLAB$ implementations in Chapter 6. The parameters that are estimated as well as realistic parameters bounds have to be deliberately selected. Those parameters are finally calibrated by modifying their bounds, such that distinct progressions of compartment sizes can be plotted by integrating the underlying system of ODEs using the estimated parameter values.

The **main purpose** of the implementations is to obtain forecasts with regard to compartment sizes during the COVID-19 spread. A built-in $MATLAB$ solver is used in the programs to integrate a system of ODEs, but a novel *nonstandard finite difference scheme* (NSFD) is alternatively applied to properly solve the respective system of ODEs. The results of the implementations including parameter calibrations are presented and analysed in Chapter 6, too.

The implementation of the first model variant is realized with respect to data of the countries Germany and Sweden. The aim is to predict future developments in compartment sizes in these European countries, which have been following different paths of dealing with the pandemic. In addition to this, a system of ODEs consisting of three distinct age groups is implemented with underlying German compartment size data. The corresponding target is the comparison between the numbers of infected and deceased individuals of different ages. A vaccinated compartment is included in the respective $MATLAB$ implementation in order to predict the effect of an increasing vaccinated fraction of the German population on compartment size progressions.

In **Chapter** 7 of this thesis, the concept of Markov-Chain-Monte-Carlo (MCMC) methods is derived as it is an alternative and stochastic approach to the modelling of infectious diseases, that is frequently applied with respect to COVID-19 in the literature. The concept of multi-state models and the approach of Bayesian inference, which are both connected to Markov-Chain-Monte-Carlo methods, are explained as well.

1.2.3   Objectives

As mentioned above, all of the predictions made in Chapter 6 of this thesis are based on $MATLAB$ implementations of systems of ODEs and the theory of compartment models in epidemiology. Summarizing, there are two major objectives of this thesis, which are characterized as follows:

- The provision of model framework conditions concerning a compartment structure, correspondent systems of ODEs, transition and transmission rates, which realistically reflect the infection-based population dynamics that are significant for the development and expansion of the COVID-19 pandemic. All used model features are derived on the basis of knowledge concerning the spread of SARS-CoV-2 and COVID-19 as well as mathematical considerations and are explained in Chapters 4 and 5.

For instance, different age groups present in populations, the effects of the modification of contact and quarantine rates as well as different vaccination strategies are incorporated into the core model.

- The week-related forecasts of the sizes of the population groups of susceptible, infected, hospitalized, deceased and recovered individuals with regard to SARS-CoV-2 and the COVID-19 pandemic.

  For this purpose, data of newly confirmed infections, COVID-19 deaths, hospitalizations, intensive care unit (ICU) admissions and recoveries are used to equip the model implementations with the necessary compartment size data. The used datasets refer to March 2020 until February 2021.

The Severe Acute Respiratory Syndrome Corona Virus Type 2

2

The second chapter of this thesis addresses the main issues that people in the whole world have been confronted with owing to the corona virus pandemic. It serves as a basis of comprehending the impact that the pandemic has on the world population. With regard to infected individuals, issues especially relate to the symptoms and course of disease of COVID-19 as well as the risk of death. With respect to affected states and societies, intervention measures which reach from distancing rules and contact restrictions to school and shop closures and usually comprise extended remote working, the extension of testing facilities, application of optimal diagnostics as well as the development and efficacy of vaccines are the currently most discussed topics. Additionally, the overload of hospitals that is caused by the severe disease progressions of a portion of COVID-19 patients, which lead to death in several cases, is a significant societal and economic issue.

Section 1.2 gives an overview of the worldwide progression of 2019-nCoV between January 2020 and March 2021, and Section 2.2 describes the aetiology and transmission paths of 2019-nCoV. Section 2.3 concerns the symptomatology and the disease process of COVID-19, Section 2.4 deals with the reasons of certain policies of containing the virus, and Section 2.5 focuses on lethality and substantially possible errors in the computation of case-fatality rates.

2.1 Worldwide Incidence of the Novel Corona Virus

The SARS-CoV-2 incidence grew faster in Europe than China in March 2020, such that Europe was called the active centre of the infection in March and April. There were 2 clearly observable peaks of SARS-CoV-2 incidence in Europe in 2020. One of them was reached in spring and the other one at the turn of the year 2020/2021. Both developed differently in different countries, which also meant 2 or more local

S. M. Treibert, *Mathematical Modelling and Nonstandard Schemes for the Corona Virus Pandemic*, BestMasters, https://doi.org/10.1007/978-3-658-35932-4_2

maxima in spring and/or autumn/winter for some countries. The attainments of the peaks together with the increasing incidence before and decreasing incidence afterwards are called the first and second wave of the pandemic.

Reacting to worldwide increasing infection numbers, the ECDC published its first risk assessment of 2019-nCoV on January 17th 2020 before the first European case was reported in France. Several further risk assessments followed during 2020 and 2021. On January 30th 2020, the World Health Organization (WHO) declared the outbreak of the novel corona virus a public health emergency of international concern. On March 11th, the Director General of the WHO declared COVID-19 a global pandemic. In June 2020, the European Commission presented a strategy to accelerate the development, manufacture and deployment of SARS-CoV-2 vaccines, of which the first ones worldwide were given in December 2020. In September 2020, the medication dexamethasone was endorsed for COVID-19 patients on oxygen or mechanical ventilation [11].

Figure 2.1 presents the daily total newly confirmed cases in Europe, North America, South America and Asia between January 28th 2020 and March 24th 2021 with a linear scaling of the y-axis (upper diagram) as well as a logarithmic scaling of the y-axis (lower diagram). Logarithmic depictions make relations within ranges of small numbers and value ranges of multiple magnitudes clearer and better manageable, such that exponential growth curves are often visualized on a logarithmic scale.

The continents Europe, North America, South America and Asia are composed of the countries that the online source Ourworldindata.org assumes [13]. For instance, Panama is the most southern country belonging to North America, and Colombia and Venezuela are the most northern countries belonging to South America. Russia is regarded as a European country and Turkey as well as the United Arab Emirates as Asian.

As can be seen from the upper diagram in Figure 2.1, the daily total incidence started increasing in Asia first, which was in late January 2020. A comparatively small peak of 4,526 Asian new infections per day was reached on February 15th. Around February 25th 2020, incidence started to conspicuously rise in Europe. On March 9th 2020, the number of daily new infections in Europe (1,820) passed the same one in Asia (1,469), whereas North America recorded a number of 86 and South America of only 9 on this day. The incidence in North America was ahead of the incidence in Europe between April 5th and 12th 2020, and then even clearer from April 21st until September 23rd. Europe attained a local maximum of 39, 041 daily new infections on April 18th, when Asia showed a number of only 11, 899 and South America 4, 367. Between May 27th and July 27th the daily incidence

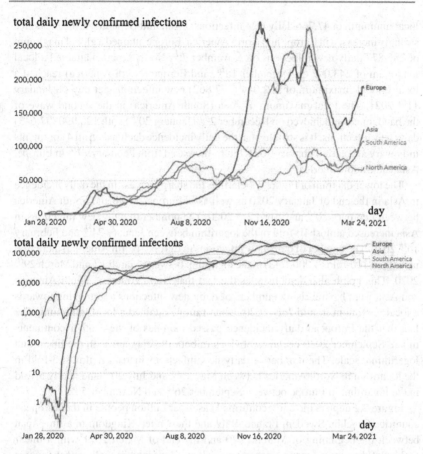

Figure 2.1 Daily total newly confirmed cases in Europe, North America, South America and Asia between the calendar week 5 in 2020 and 12 in 2021 on a linear scale (upper graph) and logarithmic scale (lower graph). Source of data: [12]

in Europe lay on a relatively low level of 12,982 to 16,267, when incidence was already rising on the other 3 continents, but then started to increase again.

North America reached a peak of 79,252 daily new infections on July 20th, when Europe attained a local minimum, and South America reached a local maximum of 75,071 daily new infections on August 15th. The Asian daily incidence was higher than the North American incidence from July 31st until October 28th, and achieved a local maximum of 124,394 on September 18th, when North America reached a local minimum of 47,766 daily new infections. From mid-September, the incidence strongly increased in Europe and North America. Europe attained a global maximum of 286,877 daily new infections on November 7th, North America attained a local maximum of 243,097 on December 18th, and Europe (North America) reached a local (global) maximum of 252,044 (277,862) new infections per day on January 11th 2021. The local maximum of Asia (South America) in the second wave of the pandemic was achieved on December 2nd (January 20th) with 128,990 (97,643) daily new infections. It is striking that the daily incidence declined on all 4 continents in January and early February 2021, but re-increased from February 17th in Europe, Asia and South America.

The lower diagram of Figure 2.1 clarifies the sharp increase in the daily incidence in Asia in the end of January 2020 as well as in Europe, North and South America between February 20th and April 2nd 2020. The number of daily new infections in Asia increased almost 10-fold in the logarithm between January 24th and February 14th 2020. The number of daily new infections in Europe (North America) increased more than 100-fold (3,000-fold) in the logarithm between March 1st and March 29th 2020. This graph also demonstrates the fact that North America, South America and Asia were having rising numbers of daily new infections and moving towards a local maximum in mid-July 2020, when Europe attained a local minimum. The fact that the European daily incidence passed the ones of the 3 other continents in late September 2020 becomes obvious in both the diagram with the linear and logarithmic scale. The number of daily new infections increased almost 3-fold in the logarithm in North America between May 16th and July 20th and nearly 5-fold in the logarithm in Europe between September 26th and November 5th.

Figure 2.2 depicts the daily confirmed cases per million people in the European countries Czechia, Sweden, France, Italy and the United Kingdom in a time span between the beginning of March 2020 and the end of March 2021 with a linear scaling of the y-axis (upper diagram) as well as a logarithmic scaling of the y-axis (lower diagram). Outlining the daily incidence per one million inhabitants instead of the total daily incidence results in a better ability to evaluate the extent to which the populations of different countries are affected by an infectious disease. In Germany, decisions concerning restrictive measures such as lockdowns or relaxations of interventions comply with the 7-day-incidence, which is the number of new infections per 100,000 inhabitants per 7 days, that is weekly updated by the German Robert-Koch Institute (RKI).

The upper graph of Figure 2.2 illustrates that the number of daily newly confirmed cases per million people firstly started increasing in Italy among the five European countries. Conspicuous but comparatively small peaks during the first wave of the pandemic, that lasted approximately until mid-May 2020, were a number of 203

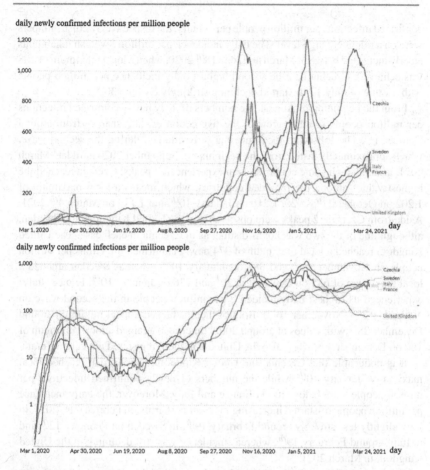

Figure 2.2 Daily newly confirmed cases per million people between the calendar week 10 in 2020 and 12 in 2021 on a linear scale (upper graph) and logarithmic scale (lower graph). Source of data: [12]

per million inhabitants on April 18th for France, 93 on March 26th for Italy, 71 on April 13th for the United Kingdom, and 24 on March 31st for Czechia. The daily incidence per million people then declined slowly in Italy, the United Kingdom and Czechia, and strongly in France, until a low level was reached in these four countries in the beginning of June 2020. While France, Italy, the United Kingdom and Czechia showed small numbers in summer, maximally attaining a value of 21 daily newly confirmed infections per million people per country, the respective Swedish numbers were on a relatively high level. The daily incidence per million Swedish inhabitants clearly increased between March and June 18th 2020, when a local maximum of 108 was achieved. Czechia had the second highest daily incidence per million people with a value of only 17. It started declining strikingly on June 30th.

From the beginning of August, the numbers of daily newly confirmed infections per million people steadily grew in the five countries, and sharply from around October 1st. The following local maxima were attained during the second wave, which approximately lasted from the beginning of September 2020 until late March 2021, and in which some countries even experienced 2 peaks: Czechia reached the highest values among the regarded countries, which means a local maximum of 1,202 on October 27th 2020, 1,210 on January 10th and 1,149 on March 4th 2021. Aside from Czechia, 2 peaks were observable in the United Kingdom and Sweden, although one of the two was not as distinct as in Czechia in each case. The United Kingdom reached a local maximum of 374 newly confirmed infections per million people on November 16th and 881 on January 10th, whereas Sweden attained a local maximum of 685 on December 22nd and 736 on January 10th. France (Italy) experienced its highest daily incidence per million people in the second wave on November 3rd (November 16th). Both of them achieved a local minimum around December 28th with values of around 200. France also attained a local minimum of 160 on December 4th, when also the United Kingdom registered a local minimum.

It is noticeable that Czechia, the United Kingdom and Sweden reached local maxima on January 10th, while the numbers of newly confirmed infections per million people were below 300 in France and Italy. Moreover, the daily incidence per million people re-started increasing in Czechia sharply on February 1st 2021, in Italy slightly less strongly around February 19th, in Sweden on February 12th and in Italy around February 19th, whereas incidence was still declining in the United Kingdom in March 2021.

The lower graph of Figure 2.2 makes clearer than the upper one that France experienced a strong decline in the daily incidence per million inhabitants between April 2nd and April 6th 2020. Additionally, it is obvious that the number of Swedish daily new infections per million people decreased by around one third in the logarithm between June 28th and July 28th, leaving the high level of April to June. What can also be better seen from the lower graph is the fact that the number of Italian daily new infections per million people increased around 20-fold in the logarithm between March 2nd and March 22nd 2020. This strong increase was followed by the other four countries around 2 weeks later. The clearly observable low level of Italian daily incidence per million people between mid-June and the end of August is also more striking in the lower graph. It reached a local minimum of less than 3 on July 3rd 2020.

Incidence in Germany in 2020

The RKI weekly publishes the German SARS-CoV-2 incidence for 19 different age groups. These data show that the first German peak of a total of 36, 094 infections of 2019-nCoV was reached in the calendar week 14, after which incidence sank to a minimum of 2, 425 infections in the calendar week 28 [14]. From the calendar week 14, the German incidence increased, which happened sharply from the calendar week 40, until a second local maximum of 138, 246 infections in total was reached in the Christmas week 52 in 2020 [14].

To create Figure 2.3, the correspondent data of the calender weeks 10 to 53 in 2020 were condensed to 7 age groups. These age groups refer to the 0–14-, 15–34-, 35–44-, 45–59-, 60–69-, 70–79- and over 80 year-old people and are expressed per 100,000 inhabitants. This aggregation of 19 to 7 age groups simplifies the subdivision of the population into even less age groups characterized by different SARS-CoV-2 infection numbers. The more exact target of this subdivision is a reduction in the number of age groups that should be considered to represent different degrees of susceptibility in the population when modelling the SARS-CoV-2 pandemic.

Figure 2.3 depicts the weekly number of newly confirmed SARS-CoV-2 infections by age group per 100, 000 people and is based on the data obtained from the RKI.

Figure 2.3 demonstrates that the age groups of over 80-, 70–79- and 60-60-year-old people attained a local maximum in the calendar week 14 with 68.94, 41.43 or 38.53 newly confirmed infections per 100,000, respectively. The age group of 10–14-year-old children reached a smaller peak of 9.21 in the calendar week 14. In contrast to that, the remaining three age groups attained a local maximum in the first wave in the calendar week 13 with values of 57.54 for the 45–59-, 54.13 for the 15–34- and 43.67 for the 35–44-year old people.

Between the calendar weeks 20 and 30 in 2020, the level of newly confirmed infections per 100,000 people was below the value 9 for all 7 age groups. The 15–34-year old individuals recorded the highest numbers during that time, with a relatively small peak of 8.94 in the calendar week 25, and the second highest level was registered for the 35–44-year-old people with a peak of only 7.63 in the calendar week 25.

From the calendar week 30, figures rose sharply for the 15–34-year old people and reached a local maximum of 28.96 in the calendar week 34. The incidence per 100,000 began to strongly increase for all of the age groups in the calendar week 40 starting on September 28th 2020.

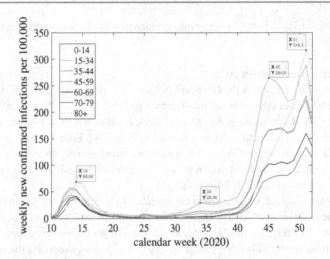

Figure 2.3 Weekly number of German newly confirmed SARS-CoV-2-infected cases by age group per 100, 000 people, for 7 age groups. Source of data: [14]

Incidence per 100,000 in the over 80-year-old people passed incidence per 100,000 in the 0–14- and 60–69-year-old people around the calendar week 43, in the 35–44- and 45–59-year-old people in the calendar week 47, and even in the 15–34-year old people in the mid of calendar week 50, until a maximum of 316.3 newly confirmed infections per 100,000 people was reached in the calendar week 51.

It is conspicuous that the incidence in the 15- to 34-year-old people clearly had 2 local maxima in the second wave in Germany, that occurred in the calendar weeks 45 with 264.5 and 51 with 288.1 newly confirmed infections per 100,000 inhabitants. As opposed to the high peaks in the over 80- and 15–34-year-old people, the 45–59-year-old people experienced a peak of 230.2 new infections per 100,000 and the 35–44-year old individuals exhibited a maximal value of 222.6 in the calendar week 51. The local maxima attained by the age group 60–69 was 159.4, so almost half of the value reached by the age group 80+. The maximal value attained by the age group 70–79 (0–14) in the second wave was 133.4 (116.9) in the calendar week 51.

Owing to the course of the incidence in the different age groups, it seems reasonable to subdivide the population into three main age groups, of which the first one comprises children and teenagers, the second one contains adults and the third one encompasses elderly persons. The comparatively high incidence per 100,000 inhabitants in the over 80-year-old people stands out clearly. This age group is nonetheless condensed with the 60–69- and 70–79-year-old people in the model realized in Chapter 5 and implemented in Chapter 6, since the absolute numbers of new infections are used as the modelling and forecast basis there, and the peaks in total incidence of the three elderly age groups in the calendar week 51 are close to each other. The peaks are 21,030 for the over 80-, 17,220 for 60–69-, 11,090 for 70–79-year old people, whereas the peak in total incidence for the 15–34- (45–59-)year old people in the calendar week 51 is 45,490 (42,080). The creation of a respective compartment model including age groups is presented in Section 5.2.4. The age group division used there is the separation of the German population into the groups of 0–19-, 20–59- and over 60-year-old people.

Outbreaks of the pandemic in Germany are characterized by their concentration on certain communities, that are affected disproportionately strongly during a specific time period [15]. Local infection herds with increased transmission activity have been traced back to events such as carnival parades, concerts or other festivities as well as ski holidays, life together in a confined space and work in the meat industry [15].

A project collects examples of transmission events from around the world, ranging from cases of individual transmission to so-called "superspreading events", provides them on a website as an interactive map and analyses the features of the settings in which they occurred [16]. A superspreading event is an event in which an infectious disease is spread much more than usual, while an unusually contagious organism infected with a disease is known as a superspreader [17]. A part of the superspreading events conform to the so-called Pareto principle or 80-20 rule, which states that approximately one fifth of all infected individuals are responsible for four fifth of transmissions [17]. Demonstrations like the so-called "lateral thinker" protests in Leipzig and Berlin in November 2020, where the protesters criticized the present corona policy to prevent the new infection protection law, could be described as superspreading events [18]. A group of researchers from the United Kingdom differentiates between "societal" superspreading events, which pose a significant threat as members of the event are free to mingle and can infect individuals in the outside community, and "isolated" superspreading events, that can be effectively quarantined as only a few individuals can transmit the virus from the isolated to the outside community [19]. Specific events concerning work in meat processing factories, religious gatherings like church services, large sports events, visiting clubs, attending weddings, funerals, conferences, shopping in malls, school attendance and hospital treatments all over the world are counted as "societal" in the study, whereas certain events in elderly care, navy or cruise shipping and prisons are considered as "isolated". It is stressed that only a few superspreading events have been traced back to air travel [19].

A 7-day seroepidemiological observational study among 919 inhabitants of the German city Gangelt, in which a superspreading carnival event had taken place shortly before, revealed that there was a positive association between celebrating carnival and the number of reported symptoms, and the number of detected infections was 5 times larger than the number of reported cases [20].

An American case-control study showed that adults with confirmed COVID-19 were approximately twice as likely as were control-participants without confirmed COVID-19 to have reported dining at a restaurant in the 14 days before becoming ill [21]. A Japanese study investigated clusters of COVID-19 cases in Japan between January and April 2020 and showed that care and healthcare facilities were the primary source of clusters, and many clusters were associated with heavy breathing in close proximity [22].

A cluster is a large aggregation of new infections within a particular geographical location or period, which can lead to fast exponential growth in the worst case [23].

2.2   Aetiology and Transmission Paths

The severe acute respiratory syndrome corona virus 2 (SARS-CoV-2) was discovered in bronchoalveolar lavage samples, that were obtained from clusters of patients who presented with pneumonia of unknown cause in Wuhan City, Hubei Province, China, in December 2019 [25]. Corona viruses are a large family of enveloped RNA viruses and others that circulate among mammals and birds [26]. It is rare that human beings are infected with them [26].

SARS-CoV-2 belongs to the sarbecovirus subgenus of the *coronaviridae* family and is the seventh corona virus known to infect humans [26]. It is a zoonotic virus, of which bats seem to be the reservoir, even though the intermediate hosts have not yet been identified [27]. Corona viruses are divided into alpha-, beta-, gamma- and delta-viruses. Many of the corona viruses known to cause disease in humans, which are alpha- or beta-corona viruses, can infect several animal species as well. The severe acute respiratory syndrome corona virus (SARS-CoV) infected civet cats, and humans in 2002, and the Middle East Respiratory Syndrome (MERS-CoV) is found in dromedary camels and infected humans in 2012 [1]. All SARS-CoV-2 isolated from humans are genetically closely related to corona viruses isolated from bat populations, and more specifically bats from the genus *rhinolophus* [1]. As a novel beta-corona virus, SARS-CoV-2 shares 79 % genome sequence identity with SARS-CoV and 50 % with MERS-CoV. Of the 4 structural genes, SARS-CoV-2 shares more than 90 % amino acid identity with SARS-CoV [28].

The receptor-binding domain (RBD) in the spike protein is the most variable part of the corona virus genome. A spike protein is a protein structure that protrudes to the outside of a virus envelope. There are 6 RBD amino acids that are critical for binding to angiotensin-converting enzyme 2 (ACE2) receptors and determining the host range of SARS-CoV-like viruses. SARS-CoV and SARS-CoV-2 use ACE2 as their cell entry receptor. ACE2 is a transmembrane protein best characterized for its homoeostatic role in counterbalancing the effects of ACE on the cardiovascular system. The carboxypeptidase activity of ACE2 converts angiotensin II to the heptapeptide angiotensin, which is a functional antagonist of angiotensin II, that is created by ACE [29]. The expression of ACE2 in the lungs of mice is reduced by a SARS-CoV infection. The novel corona virus seems to have a RBD that binds with high affinity to ACE2 from humans, ferrets, cats and other species with high receptor homology [30]. Depletion of ACE2 may be causative in the lung injury caused by SARS-CoV and SARS-CoV-2, and high plasma angiotensin II is reported in patients with COVID-19 [29]. The published genetic sequences of SARS-CoV-2 isolated from human cases are very similar and suggest that the outbreak resulted from a single point introduction in the human population around the time that the virus was primarily reported in humans in Wuhan [1]. Given the similarity of SARS-CoV-2 to bat SARS-CoV-like corona viruses, it is likely that bats serve as reservoir hosts for its progenitor, such that natural selection in an animal host before zoonotic transfer is a theory of the SARS-CoV-2 origin [30]. A second theory is natural selection in humans following zoonotic transfer, meaning that a progenitor of the virus acquired the SARS-CoV-2 specific genetic features during undetected human-to-human transmission. The SARS-CoV-2 spike protein optimized for binding to human-like ACE2 is the result of natural selection [30]. If SARS-CoV-2 pre-adapted in certain animal species, the risk of future re-emergence events would seem small. If the adaptive process occurred in humans, zoonotic transfers would be unlikely to take off without the same series of mutations [30].

SARS-CoV-2 is distinct from SARS-CoV, which firstly appeared in Southern China in 2002, and MERS-CoV, which occurred on the Arabian Peninsula in 2012 [31]. In contrast to the SARS-CoV epidemic, the SARS-CoV-2 pandemic has a lower pathogenicity but appears to be much more contagious [32]. Moreover, many patients infected with SARS-CoV-2 are reported to develop low-titer neutralizing antibody and usually suffer prolonged illness, suggesting a more effective SARS-CoV-2 immune surveillance evasion compared to SARS-CoV. The immune surveillance evasion of SARS-CoV-2 is described as more effective than the one of SARS-CoV-2 [32]. SARS-CoV-2 mortality factors were proved to be similar to those of SARS-CoV and MERS-CoV in a systematic review with a meta-analysis including 5 databases and 28 studies to compare respective death predictors [33]. All viruses, including SARS-CoV-2, change over time. Over 170, 000 variants of the virus have been sequenced by the COVID-19 Genomics UK Consortium as of January 8th 2021 [26]. Compared to SARS-CoV-2, the corona viruses HKU1, NL63, OC43 and 229E are associated with only mild symptoms [30].

The SARS-CoV-2 infection is transmitted via direct or close contact with infected people, since saliva and respiratory secretions are primarily expelled by coughing and sneezing [27, 34, 35]. This spread of the infectious agent caused by the dissemination of droplets is called a droplet transmission [36].

Transmissions were detected in prisons as for example in Hubei, Shandong, and Zhejiang in China, hospitals and in long-term living facilities, such that close proximity among people is supposed to amplify transmission [27]. Among 344 clusters involving 1, 308 cases out of a total 1, 836 reported cases in China, 78–85 % of clusters occurred within household settings, suggesting that transmission occurs during close and prolonged contact [27].

Airborne transmission is theoretically possible but not yet verified [37]. It is defined as the spread of an infectious agent caused by the dissemination of aerosols that remain infectious when suspended in air over long distances and time [36]. It can occur during aerosol-generating procedures, but is not believed to be a major driver of transmission based on available evidence [27]. Researchers from Boston stated that viruses did not have to be transmissible via air to spread quickly and spaciously and it was difficult to connect experimental data on a possible aerosol-based transmission with aerosol-based infections over long distances, especially in well-ventilated rooms [38].

Replication-competent viruses were traceable in experimentally created and with SARS-CoV-2 viruses enriched aerosols for up to 3 hours and a reduction in infectious titers was observed in one study [39]. It has to be considered that this kind of aerosol production is artificial and mechanical [34]. Viruses were proved to be released via normal speaking in dependence of the sound intensity in other studies, which indicate that the particle emission rate is correlated with vocalization amplitude [40]. Singing in groups seems to lead to transmissions, which could be droplet- or airborne-effected [34]. Corona virus-RNA-containing aerosols were also verified in room air sampled from the exhalation of patients [40].

RNA of SARS-CoV-2 has been detected in other biological samples, including urine and feces [41]. In order to be transmissible via stool viruses have to be replication-competent [42]. Fecal-oral transmission may be possible, but there is only limited circumstantial evidence to support this mode of transmission [26].

It is not yet clear to what extent smear infections occur due to contaminated surfaces, that are called fomites [37, 43]. Such contact transmissions cannot be ruled out as replication-competent SARS-CoV-2 viruses were verified in environment surfaces [34]. Transmission via breast milk has not been proved, and cases of infections of newborns were only infrequently reported [34].

Aside from SARS-CoV-2, influenza and influenza-like-illnesses (ILI) are also transmitted via airborne infection, requiring close contact to an infected case. Aerosols are a significant transmission route of influenza and ILI, but assumed as rare for 2019-nCoV [44]. Influenza is an acute respiratory virus infection effected by orthomyxoviruses. The term *"influenza-like-illness"* is a generic concept which describes respiratory infections triggered by a broad range of viral agents [44]. Unlike influenza, ILI are characterized by the absence of a severe feeling of illness, subfebrile temperatures and a slower start of disease [44]. The influenza season last from October until April in the northern hemisphere. During the last years, a shift of the seasonal maximal number of infections to January/February was registered [44]. Both influenza and SARS viruses initially affect the respiratory tract. Clinical differentiation of SARS compared to influenza or ILI is given by the almost obligatory presence of a pneumonia with radiological features as well as paraclinical changes like lymphocytopenia or LDH-increase [44]. The pathogenesis of 2019-nCoV is different from the pathogenesis of influenza, since influenza is restricted to the lungs, whereas COVID-19 destroys the endothelium, which can lead to a secondary disease in organs in the whole body [45]. COVID-19 patients were proven to become diseased with acute kidney failure, severe septic shock, lung embolisms, venous thrombosis and other illnesses [45]. Additionally, they need dialysis and insulin more often than ILI patients [45].

2.3   Disease Process and Symptomatology

From a pathophysiological perspective, COVID-19 can be defined as a complex acute lung disease with a grave damage of the alveolar epithelium and the pulmonary-vascular endothelium leading to a severe respiratory insufficiency for a proportion of patients [37]. The lung pathology in severe COVID-19 is different from the earlier pneumonitis, which reflects direct ACE2-mediated infection of type II pneumocytes, as progressive losses of epithelial-endothelial integrity, septal capillary injury, a marked neutrophil infiltration, intravascular viral antigen deposition and localized intravascular coagulation are present [29]. Subsection 2.3.1 defines the main stages of disease progression from contagion to major symptom occurrence including the latency and incubation period. Subsection 2.3.2 gives an overview of the most common symptoms of COVID-19.

2.3.1   Progression of the Corona Virus Disease 2019

Individuals infected with a virus pass the latency and incubation period. The *latency period* is defined as the time period between the exposition to the virus, entailed with the infection itself, and the beginning of the infectious period of the infected person. The RKI uses a mean latency period length of 3 days in its modelling of the corona virus pandemic [46]. The *incubation period* is the time period between the exposition to the causative agent, entailed with the infection itself, and the first development of symptoms [47]. The RKI estimates it to last an average 5–6 days [10]. An Irish review of 1,084 worldwide studies concerning the length of the incubation period of COVID-19, that altogether incorporated 45,151 infected people into the analysis and were retrospective case studies, multi-centre studies, case series, prospective cohort studies, web data mining, cross-sectional studies or individual case reports, detected that the mean length of the incubation period in all studies was 6.71 days with a median of 6 days [48]. The median time from symptom onset to laboratory confirmation decreased from 12 days (range 8–18 days) in early January to 3 days (range 1–7 days) by early February 2020 in China, and from 15 days (range 10–21) to 5 days (range 3–9) in Wuhan according to the WHO [27].

The probability that an infected case becomes sick with the disease is called the *manifestation index*. It is expressed as a percentage. Its computation depends on its exact definition, including how affected the underlying population is, and how often asymptomatic cases are tested to find out if they are really asymptomatic or pre-symptomatic [10]. An issue that is often addressed is the difference in COVID-19 manifestation between men and women. According to the RKI, men and women are affected by COVID-19 to a similarly large extent, but twice as many men as women die due to the disease [10]. A study from London, which included 107 reports concerning sex-bias in the pandemic from across the world from January until June 2020, revealed that males and females had similar numbers of infections and the male sex was associated with an increased odds of ICU admission and death [49].

The *serial interval* is defined as the average interval from the symptom development of an infected case until the symptom development of a person that has been infected by this case. The average serial interval is estimated to last around 4 days by the RKI [10]. The incubation period and the serial interval are modelled with the aid of probability distributions in Subsection 4.2.2 of this thesis.

COVID-19 is subdivided into 3 main stages: The early infection, the pulmonary manifestation and the serious hyper-inflammatory phase [37]. The proliferation of the virus in the alveolar epithelium leads to an inflammatory response in the first stage. Patients report respiratory difficulties only in the further course of the disease in addition to fatigue, fever or limb pain, that can already occur due to a hypoxaemia, that is an abnormally low level of oxygen in the blood [37]. An ongoing hyperventilation can trigger further damage of the lung, which is described as a patient-self-inflicted lung injury [37]. Primarily the upper respiratory tract of the nasopharynx is affected, although the virus infection is multiplied in the areas of the lower respiratory passages and the gastrointestinal mucosa [50]. Most of the infected individuals stabilize at this stage of disease progression and do not develop a pneumonia [37].

In a phase of strong hypoxaemia, which often results in an admission to intensive care, the lung can destabilize in order that the disease equals a severe acute respiratory distress syndrome (ARDS). COVID-19 differs from ARDS in the way that the connected vascular inflammation with thromboembolic events has a higher relevance. In a post-mortem examination that compared the lungs of 7 individuals who died from COVID-19 with the lungs of 7 people who died from ARDS secondary to influenza A, the major vascular impairment of pulmonary vessels was found to be considerably large compared to a severe ARDS disease [51].

German researchers examined the first 50 COVID-19 patients hospitalized in the university clinic Aachen. The study was induced by the severe outbreak of the pandemic in the district Heinsberg close to Aachen in mid-February 2020. The average age of the patients was 65 years, and 24 of the patients had ARDS, were treated in ICU and had to be intubated, whereas the other 26 were not in ICU. The study showed that COVID-19 ARDS patients exhibited pre-existing respiratory diseases or adiposis as well as persisting increased inflammatory markers more frequently than those without ARDS. Nonetheless, COVID-19 patients without ARDS can be hospitalized for a longer time, too, for example due to increased inflammation values or needed oxygen supply [52].

It is not exactly clear how long infectious people keep their contagiousness. It has been verified that infectiousness is highest around the time of symptom development and a remarkable number of transmissions emerging from an infected person occurs before the first symptoms develop [10]. Based on different studies, the RKI estimates that contagiosity strongly declines after 10 days if the disease progression is moderate [10]. A retrospective cross-sectional study used SARS-CoV-2 polymerase chain reaction-confirmed positive samples and suggested that infectivity of patients with a polymerase chain reaction test with a cycle threshold lower than 24 days and a symptom duration of less than 8 days may be low [53]. A review of studies conducted until April 2020 found a variation over 1–4 days of the median pre-symptomatic infectious period across studies, and showed that the estimated mean time from symptom onset to 2 negative polymerase chain reaction tests was 13.4 days. Here one study provided an approximate median infectious period for asymptomatic cases of 6.5 to 9.5 days [54]. A German study among 920 hospitals and 10,021 people sick with COVID-19, that was conducted between February and April 2020, showed that ventilation lasted approximately 13.5 days (median) [55].

Different studies implied that the time period from symptom development until hospitalization was 4 days on average, but 7 days for cases with an acute lung failure [10]. A COVID-19-caused hospital stay was reported to last 10 days (median), but 16 days (median) for people who were admitted to intensive care and 18 days (median) for ventilated individuals by the RKI [10]. According to data from the ISARIC COVID-19 database, 18 % of hospitalized patients were admitted to ICU or a high dependency unit (HDU), with a mean duration of stay of scarcely 10 days and a median of 7 days [56]. The average length of stay in ICU has been reported to be around 7 days for survivors and 8 days for non-survivors, although evidence is still limited [56].

2.3.2   Symptoms of the Corona Virus Disease 2019

The European Centre for Disease Prevention and Control (ECDC) differentiates between three groups of infected people without major symptoms: Asymptomatic cases are individuals with laboratory-confirmed SARS-CoV-2 who have not shown any clinical symptoms during a 2-week follow-up period since the last possible exposure to an index case or in the 7 days since collecting the sample which gave the first polymerase chain reaction-confirmed positive result. Pauci-symptomatic cases are people with laboratory-confirmed SARS-CoV-2 showing very mild clinical symptoms. Pre-symptomatic cases are individuals who are asymptomatic when testing positive for SARS-CoV-2, but develop COVID-19-compatible symptoms during the 7-day-period after collecting the sample which tested positive [57]. Asymptomatic and pre-symptomatic virus shedding posts a big challenge to infection control and contributes to the dark figure, which is defined as the number of unconfirmed cases worldwide or within a certain country or region.

A wide range of symptoms of COVID-19, developing 2–14 days after exposure to the virus, have been reported. The symptoms include cough, a sore throat, congestion and a running nose, shortness of breath, headache, new loss of taste or smell, fatigue, muscle of body aches, fever or chills, nausea or vomiting and diarrhoea [58]. Generally, the courses of disease are nonspecific, diverse, and vary strongly from asymptomatic processes to severe pneumonias with lung failure [34]. It should be stressed that the majority of the relatively rare asymptomatic cases most often develop symptoms in the further disease process, and the proportion of asymptomatic cases is unclear [27].

According to the WHO, 87.9 % of the infected people exhibit fever, 67.7 % a dry cough and 18.6 % are short of breath, approximately 80 % of laboratory confirmed patients have a mild to moderate disease, 13.8 % undergo a severe disease (dyspnea, respiratory frequency $\geq$30/minute, blood oxygen saturation $\leq$ 93 %, PaO2/FiO2 ratio <300, and/or lung infiltrates >50 % of the lung field within one to 2 days) and 6.1 % are critical cases (respiratory failure, septic shock, and/or multiple organ dysfunction/failure) [27]. It is notable that the absence of fever in SARS-CoV-2 infection (12.1 %) is more frequent than in SARS-CoV (1 %) and MERS-CoV (2 %) [59].

An observational study of 1,420 patients with mild or moderate disease showed that the most common symptoms were headache (70.3 %), loss of smell (70.2 %), nasal obstruction (67.8 %), cough (63.2 %), asthenia (63.3 %), myalgia (62.5 %), rhinorrhoea (60.1 %), gustatory dysfunction (54.2 %) and sore throat (52.9 %). Fever was reported by 45.4 % of the patients [60]. An analysis of data from 4,203 patients mostly from China identified fever, cough and dyspnoea (80.5 %, 58.3 % and 23.8 %, respectively) as the most common clinical symptoms, and hypertension, cardiovascular disease and diabetes (16.4 %, 12.1 % and 9.8 %, respectively) as most common comorbidities [61]. Another study among 20,133 hospitalised patients from acute care hospitals in Great Britain identified the following common symptom clusters: A respiratory symptom cluster with cough, sputum, shortness of breath and fever, a musculoskeletal symptom cluster with myalgia, joint pain, headache and fatigue, and a cluster of enteric symptoms with abdominal pain, vomiting and diarrhoea [62].

For all German cases confirmed until January 12th 2021, clinical information was available for 1, 195, 918 (62 %). 40 % of these had a clinically verified cough, 27 % fever, 28 % a running nose, 22 % a sore throat, 21 % a loss of smell or taste and 1 % a pneumonia [10].

The group with the highest risk of becoming a critical case with up to life-threatening symptoms are people aged over 60 years as well a individuals with comorbidities like diabetes, hypertension, cardiovascular disease, chronic respiratory disease and cancer [27]. A crucial risk factor is a systemic inflammatory reaction with an increase in inflammatory mediators [37]. Only a very small ratio of those aged under 19 years develop to a severe (2.5 %) or critical disease (0.2 %) according to the information of the WHO [27]. An analysis in Hong Kong showed that the presence of gastrointestinal symptoms was associated with a more severe disease course [50]. A retrospective Chinese study assessed the clinical and computer tomography (CT) characteristics of nearly 300 COVID-19 pneumonia patients from multiple hospitals and identified the baseline risk factors for clinical progression. It found out that the CT severity score was associated with inflammatory levels, older age, a higher neutrophil-to-lymphocyte ratio and CT severity scores on admission were independent predictors for progression to severe COVID-19 pneumonia. Moreover, patients who developed a bacterial co-infection were significantly older and more likely to have underlying hypertension, and had a significantly longer length of hospital stay [63].

2.4 Policies of Containing the Corona Virus Disease 2019

This section deals with approaches of containing the novel corona virus, which are usually put into practice by authorities with the aid of testing extensively, curfews and lockdowns imposed on populations. Test implementations are the typical way of diagnosing the virus as well. COVID-19 intervention measures usually conform to current and expected incidence and thus to the status of the wave of the pandemic that a certain country or region is in. Since December 2020, the virus has also been tried to be controlled by governments with the aid of vaccines that were developed and tested in clinical studies over the course of 2020. Diagnosing methods are explained in Section 2.4.1. The intervention policies of Germany and Sweden are focused on in Section 2.4.2.

An increasing number of SARS-CoV-2 vaccines have been developed, examined and reviewed all over the world in the beginning of the year 2021. They are treated in Section 2.4.3. A significant question is the effect of vaccines on viral mutations and its control of the contagiousness of vaccinated people. The most important aspects of the currently most relevant SARS-CoV-2 mutations are described in Section 2.4.4.

2.4.1 Clinical Diagnosis

Diagnosis of a SARS-CoV-2 infection, that turns a person from a suspected into a confirmed case, is realized with the aid of a direct proof using a specific testing method.

The molecular genetic verification of SARS-CoV-2 with the aid of the reverse transcriptase polymerase chain reaction (RT-PCR) using throat, sputum or nasal swabs or bronchial alveolar lavages is the diagnostic gold standard [44]. A polymerase chain reaction (PCR) test is a method to analyse a short sequence of the deoxyribonucleic acid (DNA) or ribonucleic RNA. The RT-PCR method converts viral RNA into complementary DNA, which is a DNA synthesized from RNA through the enzymatic protein reverse transcriptase, which is a RNA-dependent DNA polymerase. It reproduces the genetic material of the virus multiple times, which is called amplification, in order that the virus can be verified even if it only occurs in small quantities [64].

The accuracy of RT-PCR tests is limited. For instance, the US Food and Drug Administration recommends at least 40 cycles of amplification, after which too little genetic material of the virus might still be detectable. In a recent study by researchers at the Cleveland Clinic of 5 commonly used diagnostic tests, nearly 15 % of the results were false-negative [65]. False-negative test results cannot be excluded owing to bad specimen quality or improper transport [66]. In a systematic review that tested 957 negatively tested people again with the aid of PCR, the rate of false-negative results was between 2 and 29 % in 5 individual studies, which equals an effective sensitivity of 71 and 98 % [67]. A sensitivity of RT-PCR was shown to be 93 % for bronchoalveolar lavage, 72 % for sputum, 63 % for nasal swabs, and only 32 % for throat swabs in a study of 1070 specimens collected from 205 patients with COVID-19, who were a mean age of 44 years [68]. Accuracy is also likely to vary depending on the site and quality of a sampling, as well as the stage of disease and the degree of viral multiplication or clearance [68].

Moreover, an American study compared 9 commonly used PCR tests from China, Hongkong, the USA and Germany with each other, among which the performance depended on the viral target and the degree of dilution [66]. The E-Sarbeco test of the Charité Berlin, the HKU-ORF1 test of the University of Hongkong and the 2019-nCov-N1 test of the US Centers for Disease Control and Prevention performed best here [66].

With respect to the upper respiratory tract, nasopharynx or oropharynx smear tests can be used for diagnosing SARS-CoV-2, and bronchoalveolar lavage or sputum samples can be used with regard to the lower respiratory tract. Nasopharynx swabs represent the standard procedure of testing for 2019-nCoV according to the WHO [69]. A sensitivity of 94 to 96 % of double-sided nasopharyngeal swabs was proved by a study cited by the RKI [53]. Sputum samples are associated with a lower clinical-diagnostic sensitivity, and only a few studies concerning tests of pharyngeal lavage have been conducted, which imply an approximately as high sensitivity as nasopharynx swabs if no dilution effect, influencing rinsing technique or volume is given [69].

Rapid antigen detection proves the protein structures of SARS-CoV-2 and works like a common hCG-verifying pregnancy test. Antigen tests are also called *point of care tests* (POCT) [70]. Specimen of nasopharyngeal samples are dropped directly to a test strip, present protein components of the virus react with the strip and the strip changes its color. Antigen tests are cheaper than PCR tests and provide a result after less than 30 minutes [64]. POCT tests are less sensitive than PCR tests, which means that a larger quantity of viruses is necessary in antigen than PCR tests to verify an infection such that a false-negative result is more probable. They are also less specific than PCR tests in order that false-positive tests are more probable. A positive antigen result has to be verified with the aid of a PCR test procedure, which can be automatized but takes 4–5 hours. Nonetheless, POCT tests are used as they contribute to an expansion of testing capacities [70].

Various SARS-CoV-2 testing strategies can be identified for use in countries, or in regions within a country, and for different epidemiological situations according to the objectives of control transmission, incidence and trend monitoring, assessment of severity over time, mitigation of the impact of the infectious disease in healthcare and social-care settings, rapid identification of outbreaks in specific settings, and prevention of the (re-)introduction of the virus into regions [57]. The ECDC questions the effectiveness and cost-effectiveness of population-wide testing strategies irrespective of symptoms [57]. It suggests testing strategies based on active surveillance and early detection of all symptomatic cases, developed and adapted through ongoing assessment of the local epidemiological situation [57]. Nevertheless, the ECDC proposes that staff or patients in healthcare and social-care settings or generally vulnerable populations should be able to be tested irrespective of symptoms [57].

The current testing strategy for the German population is based on testing via PCR those people who exhibit symptoms and potentially symptomatic people who live in the same household as a confirmed case with whom they had an at least 15 minutes long contact [71]. Severe respiratory symptoms like pneumonia, bronchitis or fever, loss of the sense of smell or taste, acute respiratory symptoms and the deterioration of the clinic image with ongoing acute respiratory symptoms are criteria that definitely initiate testing [72].

Apart from this, patients, inhabitants or nurses in a hospital or retirement home, people who live in a community body like a school, emergency accommodation or correctional facility are tested via PCR if a SARS-CoV-2-infected case is confirmed in the respective institution. Furthermore, patients are PCR-tested before they are admitted or re-admitted to a medical institution. Individuals with a close contact with a confirmed case during the last 14 days can be tested as the inducement of their doctor. Furthermore, individuals with mild symptoms, who belong to a RKI-defined risk group, work in the health care system with clinical contact to others, live in an area with an increased 7-days-incidence or have had or will have close contact with many other people, can be tested. This is always a decision on a case-by-case basis made by the consulted doctor. Personnel or visitors of medical, dialysis, operation-performing institutions or medical practices, day hospitals and emergency services are allowed to receive a single free antigen test. People who return from a region that is classified as SARS-CoV-2-high risk according to the Robert-Koch Institute can be tested after at least 5 days of quarantine [71]. The number of tests conducted by countries usually concerns the number of PCR tests, which diagnose the infection with a higher sensitivity than POCT tests. Serology tests can solely inform about the presence of antibodies in the blood cells due to a prior infection [73]. Serological detection methods are not used in the early diagnostics of the novel corona virus [44].

The German virologist Christian Drosten recommended to concentrate on clusters in testing, which are created by approximately every tenth infected case, in the reaction to positive tests [74]. In terms of SARS-CoV-2 a cluster leads to an exponential increase in case numbers as the chain of infected individuals does not break here. The isolation of clusters might be meaningful in the control of the pandemic. The virologist suggested testing on infectiousness instead of infectedness by determining the virus load with the aid of a PCR test [74].

Lastly, there is no specific treatment for diseases caused by 2019-nCoV, but many of the symptoms can be treated, and supportive care for infected persons can be highly effective.

2.4.2   Governmental Measures in Sweden and Germany

Countries affected by 2019-nCoV started intervening with the aid of lockdowns in March 2020. Strong recommendations or rules related to hand hygiene, physical distancing, wearing face masks, staying home when sick and avoiding unnecessary travel became common orders.

The health ministers of the European Union and G7 states said to plan to work together more closely and intensely in order to prevent a severe outbreak in Europe on February 4th 2020. On March 8th 2020, Italy issued a decree to install strict public health measures including social distancing, starting in the most affected regions and extending these measures at national level on March 11th [11]. Spain, France and many other European countries installed similar public health measures. Most of the measures taken by countries in Europe and all over the world to fight the pandemic are NPIs, which are actions, apart from getting vaccinated and taking medicine, that people and communities can take to help slow the spread of infectious diseases [75]. NPIs were released in Germany on April 20th 2020. The aim of these measures is realizing a delay of viral spreading to allow the health care system to extend its capacities, or ideally achieve a complete stop of of viral spreading [76]. Effective medication against the virus is usually not available. Mitigation therefore tends to primarily rely on NPIs, which substantially aimed at reducing contacts between infectious and susceptible individuals during the pandemic [77]. However, there is increasing concern that NPIs could increase the number of other diseases, which mental illnesses are among, and the risk of an economic recession with unforeseeable implications [76].

Intervention Measures in Germany

The German Federal Ministry of Health and the Federal Center for Health Information summed up the most relevant information for individual protection from an infection on February 7th 2020. On February 15th 2020, when China was still 2019-nCoV hot spot, the recommendations of the Council of Health Ministers concerning travellers entering from China were directly realized. On February 27th, which was before the major outbreak of COVID-19 in Germany occurred, the German Health Ministry initiated a crisis unit to fight the pandemic. The crisis team recommended a cancellation of all events with more than 1,000 expected spectators in mid-March. An obligation to wear face masks was primarily in effect in all federal states of Germany on April 22th. In the end of March, Germany tried to prepare hospitals for a severe outbreak and expand intensive care capacities. Contact restrictions were imposed on the German population in March as well. They included the closure of restaurants, hairdressing salons and restrictions concerning sports activities.

In a scientific assessment of the spread of COVID-19 in Germany the control measures for mitigation of COVID-19 that were adopted in Germany as of April 4th 2020 were summarized, grouped and their impacts on the transmission rate are estimated [78]: The closure of schools, universities, sport clubs, the cancellation of public events and encouragement for stricter social distancing are supposed to reduce child-child contacts by 80 %, adult-adult contacts by 20 % and senior-senior by 10 %, and came fully into force on March 16th 2020. The effect of the enforcement of a remote working policy, closure of all restaurants and bars and reduced public transport services are evaluated as reducing child-child contacts by 20 %, adult-adult contacts by 50 %, senior-senior by 30 %, child-adult by 20 % and adult-senior contacts by 30 %, and were applied on a national scale from March 13th 2020. Moreover, a raised awareness due to media starting from February 25th 2020 and the cancellation of big events are assumed to reduce contacts on average by approximately 20 %. The closing of non-critical businesses and sanctioning of gatherings of more than 2 or 3 persons are assumed to decrease the contact rates by another average 30 %, starting on March 22nd 2020. Expanded testing activity started on February 25th 2020 but more significantly between March 9th and March 15th 2020, and is supposed to increase case-detection rates 10-fold.

With respect to the first lockdown in Germany, intervention measures were implemented in 3 steps: Around March 9th 2020, large public events were cancelled, around March 16th 2020, schools, childcare facilities, and many stores were closed, and on March 23rd 2020, a contact ban including the prohibition of even small public gatherings and the closure of restaurants and non-essential stores were realized [79]. In a scientific approach using Bayesian inference applied to compartment models, it was shown that a model-specific spreading rate, which was derived from confirmed SARS-CoV-2 infections, decreased from a value of 0.43 by around 40 % to a value of 0.25 around March 9th, further down to 0.15 around March 16th and to 0.09 around March 23rd [79].

Due to a decreasing incidence, the restrictive measures were lifted stepwise from May to August 2020, although distance rules were still valid in all spheres of public life. Owing to travel activities used by large parts of the German population during the summer months a test obligation for people returning from areas at risk came into effect on August 8th. In September 2020, the national academy of sciences recommended consequent interventions to prevent infections. Contact restrictions and other measures to fight the pandemic were intensified in Germany and other European countries in October and November 2020. They comprised a tourist accommodation prohibition in areas with a 7-day-incidence crossing a threshold of 50 per 100,000 inhabitants. Restaurants and shops except for those selling food and basic necessities as well as drug stores were closed, cultural and sports activities were forbidden to a large extent on October 28th 2020, and people from only 2 households were allowed to meet in public from then. In December, schools were closed in Germany, which finally meant a second lockdown, that was even aggravated in January 2021. It has to be added that decisions concerning measures like substantially schools closures lead to dissent among politicians and the population in general, and were temporarily realized differently in different federal states. [80, 81]

In order to illustrate some possible effects of less than perfect compliance with NPIs on their effectiveness in curbing the spread of infectious diseases, a scientific approach simulated a situation mimicking the status of the disease in Germany in the fall of 2020. The analysis showed that a stagnation or even reduction of case numbers could be achieved in a reasonably short time only if the measures were sufficiently effective in reducing transmission and a large proportion of the population implemented them [77]. According to the ECDC, there are concerns about the long-term sustainability of following such preventive measures in terms of population acceptance, compliance, and the potential social and economic consequences [82].

Intervention Measures in Sweden
Sweden is known to implement a special strategy ("Sonderweg") in the pandemic since public life was restricted less than in other European countries. The work of

the Swedish government to response to 2019-nCoV focuses on recommendations, voluntary cooperation and individual responsibility. However, Swedish aspirations are similar to other countries as the main aim is to reduce transmissions. The most important targets of the Swedish COVID-19 policy are the reduction of mortality and morbidity in foremost the strongly affected elderly generation, as well as the minimization of various negative consequences for individuals and society [83, 84].

The Swedish cultural divergence is related to its high value on secular-rational values and self-expression values and Sweden was rated as the most individualistic of 80 nations in 2006 [83].

Compared to other countries, where schools and playgrounds were closed over long terms from March 2020, Swedish objectives were providing the opportunity for physical activity, maintaining important social functions and meeting the needs of children. Thus Swedish upper secondary schools moved to online instruction while lower secondary school remained open [85]. A systematic Swedish review revealed that children were not the drivers of transmission of the pandemic, such that school opening was regarded as unlikely to impact the mortality rates in older people [86].

The Swedish intervention activities can also be generally characterized as bearable for a longer time period compared to strict lockdowns. The Swedish healthcare system is governed on the 3 levels of the government, which is responsible at the national level to define policy and legislation and includes the Public Health Agency of Sweden (PHA, Swedish: Folkhälsomyndigheten) and the National Board of Health and Welfare, 21 regions with responsibility for healthcare and 290 municipalities providing care for the elderly and disabled [83, 84]. A unique feature of the Swedish COVID-19 approach is that the main responsibilities in crisis management are overtaken by experts, rather than politicians [83]. The Swedish state epidemiologist Anders Tegnell stated in an interview with the multidisciplinary science journal Nature that it was overstated how unique the Swedish approach was, Sweden had a strategy and found long-term solutions keeping the distribution of infections at a decent level, and it was politically impossible to lock down whole geographical areas in Sweden [87]. In the second half of the year 2020, a growing number of Swedish researchers, including the former state epidemiologist Annika Linde, criticized the delayed reactions to the virus, which was expressed by low testing capacities [88]. Additionally, king Carl XVI of Sweden stated that the special strategy of Sweden in coping with the pandemic had failed due to high lethality rates and sharply increasing incidence [89]. The number of daily newly confirmed cases per million people in Sweden reached a maximum of 736 on January 8th 2021, which was 3 times as high as the same number in Germany at this point in time [90]. The Swedish fatality rate was higher than the rates of Germany, France, Spain and the

United States between the end of September and the beginning of December 2021, which is visualized in Section 2.5.

For estimating deaths, the PHA uses $SmiNet$, which is a reporting system and database for notifiable infectious diseases and contains all PCR-confirmed cases of COVID-19 in Sweden individual-level records [91]. The city and region Stockholm accounts for 23 % of the Swedish population and is the epicentre of the SARS-CoV-2 pandemic in Sweden. As of May 25th 2020 it accounted for 48 % of all Swedish COVID-19 deaths. The PHA reported that many families from Stockholm travelled to the Italian Alps during the Swedish winter holiday week from February 24th to March 1st 2020, which coincided with the Italian disease outbreak in the Lombardy region on February 21st [91]. Stockholm experienced an initial peak in incidence while other regions had the advantage of having a few weeks to prepare for their peak [83]. Although the same policies were applied, the lethality rate of Stockholm in the first wave was 5 times as high as the rate of the region Skåne. Skåne was the region with the third highest number of COVID-19 deaths per million inhabitants in Sweden as of February 6th 2021 [92]. Between March 2020 and February 2021, its number of deceased people per million was around two thirds of the same number in Stockholm (1,054 versus 1,585), although the number of new infections per million was 1.23 times as high as the same number in Stockholm during this time period (73,852 versus 59,616) [92].

2.4.3 The Stringency Index

The *stringency index* is a part of the Oxford COVID-19 Government Response Tracker (OxCGRT) and records the strictness of the lockdown and intervention policies during the corona virus pandemic in a country-specific way [93]. It is a composite measure based on certain response indicators, which are school closures, workplace closures, public event cancellations, gathering restrictions, closures of public transports, public information campaigns, stay home measures, internal movement restrictions, international travel controls, testing policies, contact tracing, face coverings as well as vaccination policies [94]. These indicators are rescaled to a value from zero to 100, where 100 is the strictest realization of the intervention measures. If policies vary at the subnational level, the index is shown as the response level of the strictest sub-region [95].

The courses over 53 weeks of the stringency indices of Germany and Sweden are shown in Figure 2.4.

stringency index

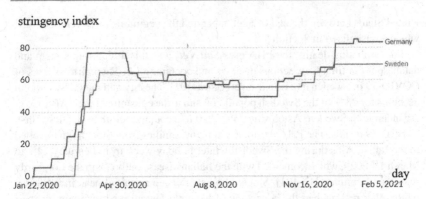

Figure 2.4 The time evolution of the stringency indices of the countries Germany and Sweden between January 22nd 2020 and February 5th 2021. Source of data: [95]

It is observable that the stringency indices of the countries change gradually. The German index starts increasing sharply from zero on January 22nd and the Swedish one even sharper on March 11th 2020, when 2019-nCoV started having a great impact on the European population. The German index reaches its first local maximum value of 76.85, which remains until May 2nd, on March 23rd 2020 and its second one of 85.19 on January 6th 2021. The Swedish index achieves its first local maximum of 64.81, that remains until June 12th, on April 5th and its second one of around 70 on December 16th 2020. The Germany index is kept on a local minimum of 49.54 between September 6th and October 14th, whereas the local minimum of 55.56 of the Swedish index lasts from August 18th until November 9th. Altogether, the German OxCGRT-stringency index fluctuates more excessively than the Swedish one. Germany responded to the global pandemic 7 weeks earlier than Sweden according to the index. It also reacted to the second major wave of SARS-CoV-2-infections 3.5 weeks earlier than Sweden.

2.4.4   Vaccines and Vaccination Strategies

The first COVID-19 vaccines were authorized in Germany in December 2020 and January 2021. The messenger ribonucleic acid (mRNA) vaccine *Comirnaty* of the BioNTech Manufacturing GmbH in collaboration with the US-American pharmaceutical company was authorized first on December 21st 2021, the mRNA vaccine *Moderna* of the US-American biotechnology enterprise Moderna in collaboration with the National Institute of Allergy and Infectious Diseases was approved on January 6th 2021 and the vector vaccine *Vaxzevria* created by the Swedish-British pharmaceutical firm AstraZeneca in collaboration with the University of Oxford was authorized on January 29th 2021 [96]. The vaccine *Vaxzevria* was named *COVID-19 Vaccine AstraZeneca* until March 24th 2021. In the past, mRNA vaccines were studied for influenza, Zika, rabies, and cytomegalovirus. As soon as the necessary information about the virus was available, scientists began designing the mRNA instructions for cells to build the unique spike protein into an mRNA vaccine [97].

The mRNA contains information from messenger RNA, including the code of a specific virus antigen and transfers the information for the production of the antigen to the protein-producing cell machinery of the body. Thus the body is able to produce the antigen on its own since cells trigger the specific immune response, which is achieved primarily through T-cells and neutralizing of the antibody production. The body is provided with the code to produce the non-infectious version of the corona virus spike protein. If a vaccinated person comes into contact with SARS-CoV-2, the immune system recognizes the surface structure and is able to eliminate the virus. The difference between conventional and mRNA vaccines is the fact that mRNA vaccines do not contain viral proteins themselves but the information of the necessity of a virus trait cell production effecting an immune response. [98]

The Moderna and Pfizer/BioNtech vaccines are delivered by lipid nanoparticles which are phospholipid membranes surrounding the mRNA that codes for the spike protein. After the injection of the lipid nanoparticle the phospholipid membrane of the nanoparticle fuses with the host membrane and releases the mRNA into the cytoplasm of the target cell. The mRNA of the spike protein is translated at the rough endoplasmic reticulum producing this protein within the cytoplasm. The spike protein is then degraded and expressed by the Major Histocompatibility Complex I (MHC I) and II (MHC II). A T-Helper cell binds with MHC II and consequently releases interleukins causing T-Helper to form T-Helper memory cells and B cells to differentiate into plasma cells, that release specific antibodies to the spike protein fragment. Cytotoxic T-cell molecules bind MHC I and a certain kind of cytotoxic T-cells bind the spike protein fragment, such that it is primed to eliminate all infected cells.

As opposed to mRNA vaccines, the vector vaccine *Vaxzevria* is composed of DNA that codes for the spike protein and is encased in a capsid from a chimpanzee adenovirus made up of a weakened version of an adenovirus of chimpanzees. The adenovirus with the accompanying DNA enters a cell by endocytosis, after which the DNA is released into the cytoplasm and migrates to the cell nucleus where it is transcribed creating mRNA that codes for the spike protein. This mRNA is translated at the rough endoplasmic reticulum, where it creates the spike protein in the cytoplasm. [99]

Comirnaty, Moderna and *Vaxzevria* authorized vaccines are given in 2 doses. A vaccination series has to be continued with the same vaccine as it has started with, with the exception of people under 60 years who have already received a first dose of the vaccine of the company AstraZeneca [100]. If different vaccines are given at first and second vaccination, this is called a heterologous vaccination. Animal data have shown that the immune response is the same after a heterologous vaccination [101]. The mRNA vaccines seem to have the advantage of their usage of the need for mRNA to enter the nucleus of the cell and potentially integrate into the host DNA. They seem to display the disadvantage to include the lower stability of mRNA as indicated by the need to store these vaccines at very low temperatures and a theoretically lower ability to stimulate the cellular manufacturing of spike protein antigen [99]. *Vaxzevria* has the advantage to be able to be shipped and stored easily at normal refrigeration temperatures, and the ability to distribute vaccines can be almost as important as the effectiveness.

Vaccine efficacy is defined as the percentage reduction in disease incidence in a vaccinated group compared to an unvaccinated group under optimal conditions, whereas vaccine effectiveness is defined as the ability of the vaccine to prevent outcomes of interest in the real world [102]. A maximum efficacy of 94.5 % against the development of clinical symptoms was verified in an US-American clinical trial with 30, 000 individuals, of whom 15,000 were vaccinated with the *Moderna* vaccine and 15,000 were given a placebo [99, 103]. The BioNtech vaccine reached an efficacy of 95 % in a clinical trial with 43, 448 individuals including a placebo group of 21,728 people and a vaccinated group of 21,720 individuals of 16 years or older [99, 104]. The *Vaxzevria* vaccine attained an efficacy of 70 % in a clinical study with 12, 000 individuals [99], and a study based on a total of 131 COVID-19 cases and 11,363 participants showed that 2 full doses of the vaccine appeared to be only 62 % effective at preventing disease, while a half dose followed by a full dose was about 90 % effective [105].

Many vaccines can prevent transmission apart from the illness itself, such that a future herd protection is possible. According to the Johns Hopkins University, it is not yet clear whether the current COVID-19 vaccines prevent the transmission of SARS-CoV-2, but it is likely they reduce the risk of virus transmission but probably not completely in everyone [106]. A study at the Maccabi Healthcare Services central laboratory in Israel showed that the viral load is reduced 4-fold for infections occurring 12–14 days after the first dose of the BioNtech vaccine [107]. An analysis based on weekly swabs obtained from volunteers in a British study showed the potential for the *Vaxzevria* vaccine to reduce asymptomatic transmission of the virus, as data showed that PCR positive readings were reduced by 67 %. The same study revealed a vaccine efficacy of 76 % after a first dose and 82.4 % with an inter-dose interval of 12 weeks [108].

Future research and clinical trials still have to address the questions how well the vaccines work in people who are at high risk of COVID-19, how well some of the vaccines protect against severe COVID-19, to what extent the vaccines prevent those who have been vaccinated from passing the virus on to others, how long immunity lasts and if the vaccines are safe for pregnant women. Apart from this, vaccine hesitancy, weariness with current public health restrictions and the staggering logistics of vaccinating the world population must be addressed by politicians and researchers [109].

The WHO tracks the development of COVID-19 candidate vaccines. 63 vaccines were in clinical and another 179 in pre-clinical development worldwide as of February 12th 2021 [110]. A heterologous recombinant adenovirus-based vaccine showed a good safety profile and induced strong humoral and cellular immune responses in participants in phase I and II clinical trials [111]. It is named *Gam-COVID-Vac (Sputnik V)* and is an adenovirus viral vector vaccine developed by the Gamaleya Research Institute of Epidemiology and Microbiology. The vaccine uses a heterologous recombinant adenovirus approach of adenovirus 26 and adenovirus 5 as vectors for the expression of the SARS-CoV-2 spike protein. It is thought to allow storage at temperatures around $-18°$ C, which is feasible for many supply chains. [112]. In a randomised, double-blind, placebo-controlled, phase III trial at 25 hospitals and polyclinics in Moscow with 21,977 adults of whom 16,501 received the vaccine and 5,476 a placebo, *Sputnik V* was associated with an efficacy of 91.6 % [111]. Another candidate vaccine is *Ad26.COV2.S* of the Janssen Pharmaceutical Companies, which is a recombinant, replication-incompetent adenovirus serotype 26 (Ad26) vector encoding a full-length and stabilized SARS-CoV-2 spike protein [113]. Serotype 26 is also used in the licensed vaccine against the Ebola virus. In a phase III study of July 2020 the vaccine candidate was 66 % effective overall in preventing moderate to severe COVID-19 among participants from different geographies in South and North America and South Africa 28 days after vaccination [114]. The definition of severe COVID-19 included laboratory-confirmed SARS-CoV-2, as well as signs consistent with severe systemic illness, admission to ICU, respiratory failure, shock, organ failure or death, among other factors [114]. Protection was generally consistent across race, age groups all variants and regions studied, including South Africa, where almost all SARS-CoV-2 cases were due to the N501Y mutation [114]. Neutralizing-antibody titers against wild-type virus were detected in 90 % or more of all participants on day 29 after the first vaccine dose and reached 100 % by day 57 with a further increase in titers in a specific cohort [113].

The ECDC suggests that the prioritization of COVID-19 vaccinations should take into account several dimensions like the objectives of vaccinating, the efficacy for different groups in the population and the protection against transmission. Vaccination of healthcare workers is described as beneficial since it improves the resilience of the healthcare system [82]. The ECDC also states that NPIs should continue to be applied if vaccines are available due to many unknowns in relation to characteristics of the vaccines, which concern deployment, supply, and to the future appearance of vaccine escape variants. In a mathematical model of the ECDC, which simulates the transmission of SARS-CoV-2 in the European Union and the progression to COVID-19, the following was assessed: If a vaccine only prevented severe disease but did not prevent transmission, a universal adult vaccination program concerning adults aged 60 years and over would prevent 90 % of the deaths, but additionally vaccinating adults aged 18–59 would account for only further 10 %. If a vaccine had 20 % efficacy against infection and an increased efficacy of 70 % against clinical disease, vaccinating 18–59-old adults would prevent 69 % of the deaths which a universal vaccination program would prevent. 43 % of the deaths would be prevented by only vaccinating adults aged 80 years and over. If younger adults with preconditions were included, 97 % of the deaths would be prevented. Moreover, the model verified that vaccination of healthcare workers would prevent only 3 % of the deaths of a universal vaccination program. [82]

Around 4.6 million vaccine doses were delivered to the German federal states until February 10th 2021. They were *Comirnaty*, *Vaxzevria* or *Moderna* doses. In the case of the respective vaccine authorization Germany will receive at least 94.1 million doses of the BioNtech/Pfizer vaccine, 50.5 million doses of *Moderna*, 56.3 million doses of *Vaxzevria*, 36.7 doses of the Janssen vaccine as well as at least 74.1 million doses of the COVID-19 mRNA vaccine *CureVac CVnCoV* and 55 million doses of the COVID-19 vaccine *Sanofi/GSK* according to the Ministry of Health [115]. The RKI separates the organization and realization of the Germany COVID-19 vaccination programme into 2 phases. In the phase Ia only small amounts of vaccine are available in relation to the population size, and vaccinations are centralized and focused on specific vulnerable groups according to the recommendations of the German Standing Committee on Immunization (STIKO), the Leopoldina and the German Ethics Board. In the phase Ib, more vaccination doses are available and exposed and vulnerable groups are vaccinated. In the phase II, vaccination is not centralized anymore such that established doctors can vaccinate people and pharmacies and wholesale traders distribute the vaccines. The general population is vaccinated in the second phase. [116]

In Germany, only individuals aged between 18 and 64 were vaccinated with *Vaxzevria* until mid-March 2021, since the STIKO stated that there were not enough reliable data available with regard to the efficacy of the vaccine for people aged over 65 [117]. Nonetheless, the WHO recommended that people aged 65 or older should be vaccinated with *Vaxzevria* earlier than younger individuals [118] and the company AstraZeneca denied insufficient vaccine effects for older people [119]. In mid-March 2021, the vaccine was publicly criticized due to a probable related increased risk of coagula and thrombocytopenia in order that several countries stopped vaccinating *Vaxzevria*. Undoubtedly, the damage of precautionarily suspending the vaccine and not using available vaccine doses would lead to an immense damage if it was not true that vaccinated individuals died more often owing to the vaccination than people who have not been vaccinated. In Germany, the Paul-Ehrlich Institute and the health ministries of the federal stated decided to resume vaccinating *Vaxzevria* on March 19th 2021 in concert with the European Medicines Agency. In the end of March, it was decided that only people aged over 60 should be vaccinated with *Vaxzevria*.

The German Ministry of Health lists over 80-year-old people, nurses and carers working in the mobile care sector, persons working in medical institutions where they are exposed to a high contagion risk, and individuals who regularly treat or look after people that are exposed to a high risk of a severe or lethal disease progression like those working in oncology or transplant medicine as those who have the highest priority to be vaccinated. People aged over 70 years, or who just received an organ transplant, are close contact persons of a pregnant woman, work in a stationary or semi-stationary institution for mentally or psychologically impaired persons or in the force of law and order with risk-exposure or in the public health service, or regularly support elderly people or people in need of care according to the social security state book, or live in refugee or homeless shelters, or have trisomy 21, dementia, a malignant haematological disease, solid tumor cancer, a severe chronic lung or liver disease, diabetes or adiposis as those with a high priority of vaccination. Over 60 year-old people belong to the population group that is classified as having an increased priority to be vaccinated. [115]

As of February 10th 2021, the states with the highest proportion of vaccinated population were Israel with 69.46 % (6,012,294 vaccinations), the United Arab Emirates with 47.37 % (4,684,658 vaccinations), the Seychelles with 45.17 % (44,423 vaccinations), the Cayman Islands with 21.43 % (14,086 vaccinations) and Great Britain with 20 % (13,577,851 vaccinations). The United States of America vaccinated 13.39 % of its population, Spain and Switzerland around 4.8 %, Italy 4.58 %, Germany 4.2 %, Sweden 3.99 %, France 3.5 % and Brazil 1.94 % up to February 10th 2021 [120]. Japan, Australia and Canada secured more than 1 billion vaccine doses altogether although only 1 % of the worldwide SARS-CoV-2 infections occurred in these countries [121]. If the 13 most promising vaccine candidates were approved,

Canada would receive more than 4 times and the European Union almost 2 times as many doses as it has inhabitants [121].

The president of the European Commission Von der Leyen stated the following positive and negative aspects concerning the distribution of vaccines in an interview with the newspaper ZEIT (volume 8 in 2021) in February 2021. She said it was positive that all 27 EU states gained access to secure vaccines. Moreover, the COVID-19 Vaccines Global Access organization (COVAX) co-led by the Coalition for Epidemic Preparedness Innovations and WHO had the aim to accelerate the development and manufacture of COVID-19 vaccines and to guarantee fair and equitable access for every country in the world, such that also poorer countries could receive vaccines. Von der Leyen stated that 85 % of worldwide states did not have access to any vaccine, it was not helpful if all Europeans were vaccinated but mutations reached Europe from for example Africa, and COVAX would start delivering vaccine batches in February 2021. She mentioned that the vaccine approval procedure had been performed thoroughly in Germany, which took long, such that the United Kingdom with its risky emergency vaccine admission was 4 weeks ahead of Germany up to then. She also saw the initial advantage of the United States in the Biomedical Advanced Research and Development Authority (BARDA), which is a central organization for pandemic crisis scenarios. Von der Leyen substantiated the currently observable vaccine shortages with the highly complicated production of mRNA vaccines and the necessity of an establishment of supply chains for them.

2.4.5 Mutations

A frequently discussed issue with respect to the novel corona virus pandemic are viral mutations. The enormous number of human-to-human transmission events has provided abundant opportunity for the selection of sequence variants [122, p. 2]. Since the circulation of SARS-CoV-2 in human beings started, the viruses have acquired an increasing number of polymorphic nucleotide positions in different frame shifts of the viral genome [123]. Thus the viruses can be separated into distinct classes. It has not been conclusively elucidated in how far specific mutations affect virulence, transmissibility or immunogenicity.

The spike protein helps SARS-CoV-2 docking on cells by binding a receptor called ACE2 on the surface of human cells in order that the absorption into and infestation of the cell are effected [124]. A worldwide growing number of virus variants exhibits the spike mutation D614G, which implies an increased transmissibility [123]. Additionally, D614G favours the binding of ACE2 on the target cells, which results in a higher infectiousness [122, p. 6]. The investigation of D614G identified that the variant provided a selective advantage through increased cellular infectivity, but there was no identifiable effect on infection severity or outcome [125]. A study with transgenic mice and Syrian hamsters showed that the D614G substitution enhanced SARS-CoV-2 infectivity, competitive fitness, and transmission in primary human cells and animal models [126]. The WHO regularly assesses if SARS-CoV-2 variants effect changes in transmissibility, clinical presentation and severity, or if they impact on countermeasures, including diagnostics, therapeutics and vaccines. According to the WHO, the SARS-CoV-2 virus with the D614G substitution does not cause more severe illness or alter the effectiveness of existing laboratory diagnostics, therapeutics, vaccines, or public health preventive measures [127].

The recently detected SARS-CoV-2 lineages B.1.1.7, B.1.351, and B.1.1.28 are of greatest concern. Each of them possesses a unique constellation of mutations, but all exhibit the N501Y mutation, which is a spike protein polymorphism. Since mid-December 2020, the United Kingdom has reported a growing number of viruses of B.1.1.7, that have spread in the south of the United Kingdom since September 2020. These viruses are characterized by an exceptionally high number of non-synonymous polymorphisms in the spike protein [123]. Non-synonymous single nucleotide polymorphisms lead to a change in amino acids in the affected codon, that is a variation pattern in the sequence of the nucleobases of the mRNA. More exactly, the B.1.1.7 variant has 23 nucleotide substitutions and is not phylogenetically related to the SARS-CoV-2 virus circulating in the United Kingdom at the time the variant was detected [127]. A simpler transmissibility of the B.1.1.7 variant is assumed by now owing to contact tracing data of Public Health England [123, 127]. A British study estimated the transmissibility of the variant at 56% higher than pre-existing variants on the basis of publicly available data from the COVID-19 Genomics UK Consortium [128]. A study investigated SARS-CoV-2 pseudo viruses bearing either the Wuhan reference strain or the B.1.1.7 lineage spike protein with sera of 16 participants in a previously reported trial with the *Comirnaty* vaccine and found equivalent neutralizing titers to both variants among the immune sera, such that it is unlikely that the variant viruses will escape the vaccine [129]. Scientists from the United Kingdom reported evidence suggesting the B.1.1.7 variant might be associated with an increased risk of death compared with other variants [131]. Another study that analysed a dataset linking 2,245,263 positive SARS-CoV-2 community tests and 17,452 COVID-19 deaths in England from September 1st 2020 to February 14th 2021 confounded an increase in COVID-19 mortality and more severe illness associated with the lineage B.1.1.7 [130].

In mid-November, routine sequencing by South African health authorities found out that a new SARS-CoV-2 variant had largely replaced other SARS-CoV-2 viruses circulating in the Eastern Cape, Western Cape, and KwaZulu-Natal provinces [127]. In December 2020, an increased occurrence of the virus variant B.1.351 was reported in South Africa, exhibiting multiple non-synonymous mutations in the spike protein [123]. Preliminary studies suggested that B.1.351 was associated with a higher viral load, which might imply a potential for an increased transmissibility [127]. Two South African preprints showed that the variant was less susceptible to neutralization by convalescent sera from individuals exposed to earlier variants, in either live virus or pseudo virus neutralisation assays [132]. Some evidence indicated that one of the spike protein mutations, E484K, might affect neutralization by some polyclonal and monoclonal antibodies [131]. The mutation E484K is a mutation through which the amino acid glutamine is replaced by lysine at the position 484 of the spike protein [133]. Public Health England reported that the COVID-19 Consortium had identified the E484K mutation in 11 samples carrying the variant B.1.1.7 after analysing 214,159 sequences [133]. Up to the beginning of February 2021, the United Kingdom had also detected 105 cases of the South African variant B.1.351 [133].

The variant B.1.1.28 circulates substantially in the Brazilian state Amazon. It also exhibits spike protein mutations [123]. A study saw potential in the association of B.1.1.28 with an increase in transmissibility or propensity for re-infection of individuals [134]. The variant B.1.1.7 was reported in the United States at the end of December 2020 and the variants B.1.1.28 and B.1.351 at the end of January 2021 [131]. In Germany, infections with the variants B.1.1.7 and B.1.351 were confirmed until January 2021 [135].

The concept of herd immunity states that only a proportion of a population needs to be immune to an infectious agent for it to stop generating large outbreaks. More exactly, herd immunity occurs when a high percentage of the community is immune to a disease usually through vaccination and/or prior illness, such that the spread of this disease from person to person is unlikely [136]. An effective vaccine presents the safest way to reach herd immunity [137]. Vaccination of 60 to 70 % of the population is necessary to achieve herd immunity according to the WHO [138]. Aside from that, the WHO regards attempts to reach herd immunity through exposing people to a virus as scientifically problematic and unethical [139]. It seems impossible to know how much of a population is immune and how long that immunity last as COVID-19 and SARS-CoV-2 have to be researched in increasing detail. These challenges could preclude plans that try to increase immunity within a population by allowing people to get infected.

It is still unknown how efficient or effective vaccines are with respect to known mutations of 2019-nCoV. According to the Johns Hopkins Bloomberg School of Public Health, SARS-CoV-2 could still infect children before they can be vaccinated or adults after their immunity wanes, but it is unlikely in the long term to have an explosive spread because much of the population will be immune in the future [140]. Obviously, the N501Y variant contains mutations that blunt the effects of neutralizing antibodies that recognize the receptor-binding and specific domains of the spike protein [141]. Furthermore, BioNTech researchers revealed that the spike mutations in B.1.1.7 had little effect on sera from 16 people who had received the vaccine of the company developed with Pfizer. Meanwhile, a team of the University of Cambridge proved that the sera of 10 out of 15 people was less effective against B.1.1.7 than against other versions of SARS-CoV-2 [141]. A study of the Columbia University showed that the serum of 22 people vaccinated with *Moderna* or *Comirnaty* was 6–9 times less potent against B.1.351, and serum from 20 previously infected people was 11–33 times less potent [142]. These and similar studies indicate a decreased effect of currently available vaccines against more easily transmissible variants of the novel corona virus.

Nonetheless, researchers are still debating if the new variants could undercut the effectiveness of these first-generation vaccines [141]. An US-American study detected no significant impact on neutralization against the B.1.1.7 lineage of the corona virus, but a reduced neutralization against the mutations present in the B.1.351 lineage, also called the N501Y variant [143]. Another study discovered that the activity of mRNA-based vaccines against SARS-CoV-2 variants encoding E484K or N501Y or the K417N:E484K:N501Y combination was reduced by a small but significant margin [144]. In a third study, the South-African N501Y variant showed substantial or complete escape from neutralizing antibodies in COVID-19 convalescent plasma, which might highlight the prospect of re-infection with antigenically distinct variants and a reduced efficacy of current spike-based vaccines. [145].

2.5   Lethality in Corona Virus Disease 2019 Cases

Whereas lethality is the proportion of deceased confirmed cases among all infected cases in a population concerning a certain infectious disease within a certain time period [146], mortality is the proportion of deceased confirmed cases among all individuals in a population. The *case-fatality rate* (CFR) is a measure to represent the lethality of a disease. In order to compute the CFR of the novel corona virus with reference to a certain population, the number of confirmed deceased cases is divided by the number of confirmed infected cases. The rate can be related to a specific time period. Hence, the CFR denotes the fraction of confirmed infections that ends in death [147, p. 32]. The difference of the lethality rate to the CFR of an infectious disease is the fact that the divisor of the lethality rate is more general and can still be specified. Using the CFR instead of other lethality rates is reasoned by the fact that the number of all infected people is unknown [10], such that usually only detected cases can be counted as infected cases. The CFR is often used as a measure of the virulence of a virus [148], which is defined as the degree of pathogenicity of the infectious disease within a group or species under controlled conditions [149].

2.5.1   Case-fatality in Corona Virus Disease 2019 Cases

No reliable data are available concerning the number of individuals infected with SARS-CoV-2. The number of cases that have been tested and are therefore confirmed is often used as the divisor in the CFR. In general, the CFR is based on diagnosed cases of disease rather than the number of actual infections [148]. For instance, the RKI counts all deaths that are associated with COVID-19, which includes deaths that are potentially not attributable to the disease [150]. The WHO relies on the fatality numbers reported by national governments while the Johns Hopkins University also takes into consideration press reports on case fatalities [6].

In the sequel, the term deaths related to SARS-CoV-2 will refer to the deaths counted as SARS-CoV-2 deaths in a specific country, that have not necessarily been caused exclusively or primarily by 2019-nCoV. According to the Johns Hopkins University, the twenty countries that were most affected by 2019-nCoV in terms of the highest absolute daily numbers of deaths by January 2021 were Mexico, Italy, Hungary, Indonesia, South Africa, the United Kingdom, Argentina, Colombia, Canada, Brazil, Spain, France, Poland, Germany, Russia, Ukraine, the United States, Czechia, India and Turkey [151]. The sequence of enumeration reflects the order of sizes of the computed case-fatality rates of these countries in a descending way. The CFR of Mexico was stated as 8.7 %, of Italy as 3.5 %, of Germany as 2.2 %, of the United States as 1.7 % and of Turkey as 1 % on January 12th 2020 [151].

In the following diagram, the progression of the case-fatality rates of three European countries, the European Union on average and in the United States over the first and second wave of the pandemic are illustrated on a linear scale.

It is conspicuous in Figure 2.5 that the case-fatality rate of the United States decreases until March 23rd 2020 while the rates of the other regarded countries increase. Among the three European countries, the CFR of Italy starts increasing at first, namely on March 2nd, whereas the rates of Sweden and Germany begin to grow about one week later, whereby the CFR of Sweden increases a lot sharper than the CFR of Germany. The Swedish CFR reaches a first peak of 9.34 % on April 3rd and an even higher one of 12.18 % on May 8th. The CFR of the United States

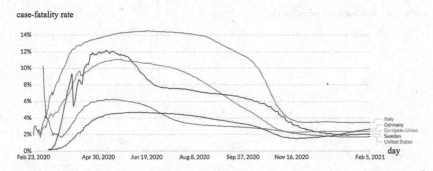

Figure 2.5 SARS-CoV-2 case-fatality rates of Germany, Sweden, Italy, the European Union and the United States between the calendar weeks 8 in 2020 and 6 in 2021. Source of data: [152]

attains a lower local maximum of 4.52 % on May 14th, and the CFR of Germany remains on its highest level between 4.5 % and 4.7 % between May 13th and July 5th. Moreover, it is evident that the German CFR passes the one of the United States on July 5th and remains on a higher level until October 15th 2020.

The CFR of the European Union is higher than the one of the United States from March 13th 2020, shortly achieving an equal level as the United States on November 7th. It stands out that the CFR of Italy is significantly higher than the rates of the other presented countries as well as the European Union on average between March 1st and October 31st 2020, with 1.81 more percentage points than Sweden on May 9th and 3.25 more percentage points than the European Union on June 2nd. The Italian CFR remains higher than the other presented CFRs from November 1st, remaining at around 1.1 percentage points over the CFR of the European Union from December 10th on.

Figure 2.6 shows the daily newly confirmed deaths per million people between March 2020 and August 2020 in the upper diagram and between September 2020 and February 2021 in the lower diagram in four European countries, on average in the European Union and in the United States on a linear scale.

The upper diagram in Figure 2.6 presents that figures sharply increased in Italy at first, which was in the end of February 2020. In France, the increase was even stronger than in Italy, started in the first week of March up to a peak of 14.94 attained on April 9th. In the second week of March, figures began to strongly increase in Germany until a peak of 2.97 was reached on April 21st and Sweden until a local maximum of 10.64 was attained on April 24th. Obviously, the Swedish peak was 3.58 times and the French 5.03 times higher than the German one. The Italian

daily newly confirmed deaths per million people

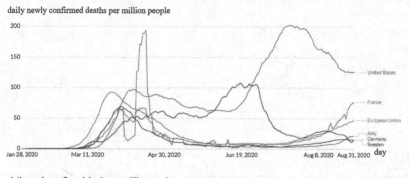

daily newly confirmed deaths per million people

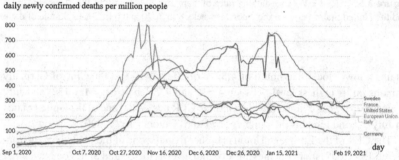

Figure 2.6 Daily newly confirmed deaths related to SARS-CoV-2 per million people in Germany, Italy, France and Sweden, the European Union and the United States between the calendar weeks 5 in 2020 and 7 in 2021.Source of data: [90]

turning point of spring occurred at a value of 13.47 on April 2nd, which was one week before the French peak. In contrast to Italy, France and Sweden, the German local maximum of April was below the maximum of spring of the average of the European Union. In all four depicted countries as well as the European Union in general, the number of daily newly confirmed deaths per million inhabitants was low between June and September 2020.

In all of the countries and the European Union on average, 2 more or less clear local maxima are observable in the second wave of the pandemic. This can be seen from the lower graph of Figure 2.6. Sweden reached a value of 7.62 on December 17th and 13.62 on January 20th, Germany achieved a value of 7.90 on December 3rd and 10.68 on January 13th, France attained a value of 9.59 on November 20th and 7.03 on February 11th, in Italy the values of 12.26 on December 3rd and 8.40 on January 11th can be seen, and the United States exhibit a value of 8.56 on December

22nd and 10.12 on January 13th. This verifies that the first peaks of the second wave were reached in France and Italy, but shortly before Christmas 2020 in Germany, Sweden and the United States. It catches attention that the maximally achieved value of autumn/winter is 1.28 times larger than the one of spring 2020 in Sweden, whereas the maximum reached in Italy (France) is 1.10 times (1.55 times) larger in spring than in autumn/winter.

What is most striking in the lower graph is the fact that the maximal number of German daily newly confirmed cases per million people attained in autumn/winter (10.68, January 13th 2021) was more than 3.5 times higher than the maximal number reached in spring (2.97, April 21st 2020). In contrast to that, the local maxima in spring and autumn/winter of the European Union (7.04 on April 7th and 7.90 in mid-December) were almost the same.

The number of daily new deaths in Italy, Germany or Sweden passed the average number of the European Union on November 6th, December 21st or January 5th, respectively. It must be added that Italian death numbers were below those of the European Union from the end of August until the end of October, and then were on the same level as those of the European Union until the beginning of November. As opposed to this, French numbers already passed the average of the European Union in mid-September.

The United States showed their peak of 6.83 in the first wave 17 days after the peak of the European Union (7.04). The local maximum of the United States (10.12, January 13th) in the whole autumn/winter wave was 1.26 times higher than the one of the European Union (8.03, November 29th).

A country with higher daily death numbers per million people than all of the four depicted countries from the beginning of September until mid-November and from mid-December on is Czechia, that is not depicted in Figure 2.6, with the local maxima of 19.58 attained on November 5th and 16.50 reached on January 15th[90].

In Table 2.1, the changes in lethality related to SARS-CoV-2 infection in seven European countries and the United States can be seen. The data were created by Brunner et al. The change in lethality between a month and the previous month was computed as the ratio of additional deceased to additional newly infected individuals. The largest monthwise change per country is printed in bold.

Table 2.1 conveys that Italy and France were the countries in which the monthly computed lethality rates increased at first among the eight countries, which was in February 2020. The French lethality rate increased 1.68-times as much as the Italian one between the end of January and February. Among the six other countries, the increase in lethality related to SARS-CoV-2 between the end of February and March was largest in Spain and second largest in Sweden. The Spanish and Swedish country-specific lethality increase rates of March were higher than the French but

Table 2.1 Change in the lethality rates of Germany, France, Sweden, Denmark, Spain, Italy, the United Kingdom and the United States per month between January 2020 and January 2021

Month	Germany	France	Sweden	Denmark	Spain	Italy	UK	US
January 2020	0	0	0	0	0	0	0	0
February 2020	0	4	0	0	0	2.37	0	0
March 2020	1	6.71	7.51	2.96	8.83	11.83	3.63	1.67
April 2020	6.21	17.57	**15.74**	**5.73**	**13.68**	15.59	**18.12**	**6.79**
May 2020	**9.16**	**19.78**	10.45	4.85	9.92	**19.79**	13.66	5.76
June 2020	4.25	7.99	2.77	2.82	12.54	17.83	10.67	2.72
July 2020	1.05	1.74	2.74	0.94	0.29	5.33	4.06	1.33
August 2020	0.45	0.42	0.58	0.27	0.37	1.59	0.95	2.08
September 2020	0.4	0.47	1	0.23	0.88	0.9	0.55	1.94
October 2020	0.41	0.6	0.14	0.39	0.98	0.75	0.79	1.24
November 2020	1.15	1.86	0.63	0.34	1.99	1.84	0.61	0.85
December 2020	2.48	2.96	1.05	0.55	2.06	3.68	1.75	1.2
January 2021	5.03	1.98	2.21	2.36	0.92	3.22	2.46	1.54

Source of data: H. Brunner, personal communication, February 18th 2021

smaller than the Italian increase rate of March. Apart from this, the initial Spanish 8.83-fold increase from February to March was around 3-9 times as high as the initial increases in Germany (1), Denmark (2.96), the United Kingdom (3.63) or the United States (1.67).

In Sweden, Denmark, Spain, the United Kingdom and the United States, the largest country-specific change in lethality rates occurred in April 2020, whereas it occurred in Germany, France and Italy in May. When comparing the country-specific peaks depicted in Table 2.1, the order of countries in a descending order of peak heights is Italy, France, the United Kingdom, Sweden, Spain, Germany, the United States, Denmark. The maximal change rate reached in France (19.78) is smaller than the Italian one by only 0.1. The height of the Swedish (German) maximal change rate is around three quarters (only half) as large as the Italian one, while the Danish one is only around one quarter as large as the Italian one.

In Germany, France, Sweden and the United Kingdom the monthwise lethality rates still grew compared to the respective previous month until July 2020 but with a downward trend. Then they decreased until October (Germany, France) or November (Sweden, the United Kingdom). In Denmark and Spain declines in monthly computed lethality primarily occurred in July. They lasted until October in Spain, and until December in Denmark. Spain showed increasing monthwise lethality ratios in April, May and June 2020 with change ratios of 13.68, 9.92 or 12.54, but a sudden sharp decrease in lethality in July with a change ratio below 1, that was approximately 2.3 % of the change ratio of June. This represents the largest difference between 2 successive non-zero table entries of a country.

The United States recorded a decrease in the monthwise lethality rate exclusively in November, and Italy in September and October. While Germany, France, Sweden, Denmark, Italy and the United Kingdom registered a decreasing growth in monthly computed lethality rates between their respective peak and the attainment of a diminution of lethality rates, Spain and the United States showed fluctuating growth rates between April and June or April and October, respectively. In January 2021, Germany recorded the largest change ratio (5.03) among the eight countries, that was around 1.55 times larger than the Italian one (second largest one) and 3.25 times as large as the American one (smallest growing one). Spain was the only country exhibiting a decreasing lethality change ratio then.

SARS-CoV-2 lethality data are often visualized age-specifically [10]. According to the Robert-Koch Institute the CFR of individuals younger than around 50 years is lower that 0.1 % and the CFR of those who are older than 80 years is often higher than 10 % [10]. According to the ECDC, there is an increased risk of hospitalisation, ICU admission and death with increased age and for those with certain underlying conditions [82].

The following diagrams show the pooled weekly death data for two age groups in the data-providing partner countries of EuroMOMO between the ends of the years 2019 and 2020. EuroMOMO is a project for the continuous and near-term European detection and measurement of excess deaths [153]. Its current partner countries (as of January 2021) are Austria, Belgium, Cyprus, Denmark, Estonia, Finland, France, Germany, Greece, Hungary, Ireland, Israel, Italy, Luxembourg, Malta, the Netherlands, Norway, Portugal, Slovenia, Spain, Sweden, Switzerland, Ukraine and the United Kingdom. The red dashed lines represent a substantial increase, the blue dashed lines symbolize the baseline progress, the blue shaded domains reflect the normal range, and the yellow domains signify time intervals in which data were corrected for delay in registration. The normal blue lines stand for

the numbers of pooled death of the respective age groups. The age groups of 15–44-
and over 65-year old people were selected owing to the wide range of individuals
they comprise as well as the clear differences in the two correspondent graphs.

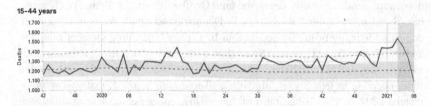

Figure 2.7 Pooled weekly death number data in the data-providing partner countries of
EuroMOMO between the calendar week 42 in 2019 and 6 in 2021 for people aged between
15 and 44 years. Source of data: [153]

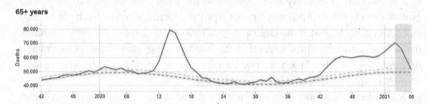

Figure 2.8 Pooled weekly death number data in the data-providing partner countries of
EuroMOMO between the calendar week 42 in 2019 and 6 in 2021 for people aged over 65
years. Source of data: [153]

Generally, the level of death numbers in Figure 2.8 is a lot higher than the level of
death numbers in Figure 2.7, which is justified by the greater mortality risk of people
over 65 years compared to 15 to 44-year old people. The highest level reached in
Figure 2.7 is 1,543 weekly deaths (calendar week 3 in 2021), which is only 3.74 %
of the lowest level achieved in Figure 2.8 (41,224 in the calendar week 25).

The progression of the graph of the weekly total number of deaths in the data-
providing EuroMOMO partner countries referring to the age group 65+ fluctuates
less than the one of the age group 15–44, such that more local maxima are found in
Figure 2.7. Nonetheless, the largest peak in Figure 2.7 is only 1.066 times as high
as the second highest (reached in the calendar week 15 in 2020), 1.097 times as
high as the third largest peak and 1.1 times as high as the fourth largest peak. The

global maximum in Figure 2.8, which is 79,432 weekly confirmed deaths of people over 65 (reached in the calendar week 14 in 2020), is also only 1.12 times as high as the second highest local maximum (reached in the calendar week 3 in 2021). As opposed to this, it is 1.48 (1.72) times as high as the third (fourth) largest peak attained by this age group.

What is most striking in the two figures is the fact that the calendar weeks 14/15 in 2020 and 3 in 2021 were the ones with clear peaks, particularly in the age group 65+. This is probably reasoned by raised numbers of deaths effected by the corona virus pandemic since the calendar week 14/15 in 2020 or 3 in 2021 lies around the peak of the first or second wave of the worldwide pandemic, respectively.

The substantial increase from the baseline value is 48,990 in the calendar week 14, 48,467 in the calendar week 15 in 2020 and even 52,513 in the calendar week 3 in 2021 for people of 65 or more years.

2.5.2 Uncertainty in Case-fatality Computations

It is generally problematic to precisely calculate the CFR because of uncertainties regarding the time course of the disease and difficulties in the diagnosis and report-ing of cases [148]. The method of time-shifted distribution analysis is a way of predicting the CFR as well as the delay between reporting and deaths over time, using only publicly available data on cases and deaths and not making assumptions or parametrisations regarding the progress of the disease [148]. Time-shifted distri-bution means that the shape of the evolving time distribution of SARS-CoV-2 cases in a given country often closely matches the shape of the correspondent distribution of deaths of SARS-CoV-2 cases, but shifted by a number of days and linearly scaled in magnitude [148].

There are many possible reasons for the fact that case-fatality rates range from 0 to more than 20 % on the country level [154]. For example, the pathogen and the immune response of the host, which can depend on age, sex, genetic factors and pre-existing medical conditions, have an impact on the SARS-CoV-2-specific CFR. Environmental factors such as climate and health system may also affect it [148]. In the following, 4 other main causes of local or global differences in case-fatality rates are given.

Firstly, the **absence of effective treatment** in certain countries may be a cause of a higher CFR compared to countries with better treatment options. This absence can be reasoned by overwhelmed health care systems effected by sharply increasing incidence rates. Results of an incidence reaching a local peak can be a deficiency of hospital beds and a reduced amount of care any individual patient can receive.

During the first wave, television companies broadcasted triage scenarios from Italy in which doctors had to decide which patients will be ventilated and which not [155]. There was also concern that the number of ICU beds and nurses in hospitals was sufficient for the volume of diseased cases in Germany [155] and other countries. In December 2020, an average 17 % of German intensive care unit beds, which meant 4, 158 beds, were not occupied according to the Federal Statistical Office of Germany. Moreover, 56.7 % of hospitalized patients had to be given artificial respiration during anaesthesia and in the intensive care unit [156].

It is difficult to find reliable data that quantifies the overburdening of the health care system. Is also has to be noted that an exorbitant ICU occupancy during pandemic times may be additionally effected by ICU cases whose ICU admittance is not necessary, although it is difficult to decide about this in marginal cases. Nevertheless, the predicted number of infected individuals at its peak can be used as a proxy for the expected demand for health care resources at the height of the epidemic [78].

Secondly, countries probably differ in the **accuracy in detection** of infections as well as the control of the observance of imposed restrictions. Different populations might follow governmental instructions in a more or less disciplined way, such that social distancing and isolation might be realized to larger measure in countries with autocratic regimes [154]. With the aid of precise mathematical models numbers of unreported cases can be determined under certain considerations, such that the dark figures of countries or regions can be computed. The German Fraunhofer Institute for Industrial Mathematics created a mathematical approach taking into account socially-weighted prevalence, which substantially means age groups and testing policies. A fundamental idea behind this was the presumption that older people were overrepresented in the number of reported infections in certain countries and asymptomatic cases or cases with only mild symptoms increased the underestimation of incidence [157]. Based on the data of April 22nd 2020, this model determined at least 22, 107 undetected infections in addition to the reported number of 73, 268 infections for Germany, and 476, 297 undiscovered infections in addition to the 105, 630 reported ones in Italy [157].

Thirdly, **testing policies** have an immense impact on the differences in casefatality rates between different countries. If a larger fraction of the population is tested, more infected cases are detected, reported and thus confirmed. Hence with more testing, more people with milder cases are identified [151]. On the one hand, this may be thought to lower computed CFRs as it can be assumed that positively tested individuals are treated earlier. It may be assumed that earlier detection due to enhanced testing enables better care for high risk patients and results in lower mortality rates [78]. On the other hand, more testing increases the number of positively tested individuals, such that more deceased cases are probably classified as SARS-

CoV-2 deaths, although their actual cause of death might be a comorbidity. Hence punctual state-funded testing facilities in certain regions or generally better testing strategies of countries could lead to a stronger overestimation and an imbalance in CFR computations between different regions.

Test prevalences are commonly translated to the overall population without incorporation of the respective dark figure of unconfirmed infected cases [158]. In December 2020, the RKI conjectured the number of unconfirmed German SARS-CoV-2 cases as 4–6 times as high as the number of confirmed ones [159]. Seroprevalence studies implied a dark figure between 2–6 times the number of confirmed infections [158]. An analysis investigated time-dependent variations in effective fatality rates over the course of the COVID-19 pandemic and used German COVID-19 surveillance data as well as age-specific fatality rate estimates from multiple international studies. Applying 3 distinct methods for estimation the analysis showed that a large fraction of time-dependent variability in case-fatality could be explained by changes in the age distribution of infections [160]. Thus the age distribution of infections was detected to be a major determinant for the resulting mortality. A Spanish analysis estimated the diagnostic rate of European countries in an unbiased way, developed an effective index to monitor the comparative situation of COVID-19 in Europe, and conveyed that the reporting rate differed between countries an regions but was roughly constant over time [161]. According to a study of 565 Japanese people evacuated from Wuhan, all of whom were tested, only 9.2 % of infected people were detected with currently used symptom-oriented monitoring [162], indicating that the number of infected people was likely to be about 10 times greater than the number of registered cases in Japan.

In a German scientific paper covering the necessity of a change of strategy in dealing with the pandemic, the so-called *notification index* NI was developed and defined [158]. It describes the dynamics of the development of the infectious disease on a national or regional level. It involves the reporting rate M_r per $100,000$ inhabitants, the rate of positive tests T_1 among the realized tests, the testing frequency T_2 and a heterogeneity factor Het. The factor Het is defined as the quotient of the percentage number of people infected in clusters to the percentage number of people infected sporadically. Its mathematical definition reads

$$NI = \frac{M_r \cdot T_1}{T_2 \cdot Het}.$$

This index is a measure for the infection risk of a region and enables the appropriate compensation of bias in dark figure estimations for regions. A larger heterogeneity factor Het implies a lower risk. The reporting rate M_r is the rate at which infected cases are reported to authorities. It could also be called the *case-confirmation rate*. The notification index was presented as an alternative to the 7-day-incidence [158]. It always has to be kept in mind that completely reliable quantitative measures are not attainable if the simple reporting rate is used, but the above notification index NI is secured by other factors. The incorporation of the reporting rate into lethality measures cannot be stopped since reliable data concerning numbers of unregistered cases do not exist. The also developed so-called *hospitalization index HI* describes the strain on the health care system and is defined as the product of the notification index NI and the rate of hospitalization K [158]:

$$HI = NI \cdot K.$$

In the computation of the hospitalization index, the parameters used for computing the index NI should be 5 days backdated as hospitalization can take place some days after infection, and the heterogeneity factor Het may be omitted [158].

A fourth factor is **age and comorbidities** like cardiovascular diseases, cancer, diabetes mellitus, chronic lung diseases, which all individually increase a mortality risk. As a result, countries with a greater share of elderly in the population, or with higher incidence of recognized comorbidity factors might be distinguished by relatively high CFRs [154]. Generally, a large factor triggering misestimation of CFRs are wrong assessments of deceased multimorbid people who were confirmed SARS-CoV-2 cases. It is difficult to decide whether they died of the infection or merely with the infection [163].

Multiple comorbidities are associated with the severity of disease progression. Many deaths and ICU admissions due to COVID-19 have been related to cardio-vascular comorbidities. A study including 1,590 laboratory confirmed hospitalized patients from 575 hospitals in 31 provinces across mainland China showed that 399 (25.2 %) reported at least one comorbidity and the most prevalent comorbidity was hypertension (16.9 %) [164]. A meta-analysis including 1, 558 infected patients in China identified hypertension, diabetes, chronic obstructive pulmonary disease (COPD), cardiovascular disease as well as cerebrovascular disease as significant risk factors for COVID-19 patients [165]. In a study including 166 UK hospitals and 16, 749 people with a confirmed SARS-CoV-2 infection, that was conducted between February 6th and April 18th 2020, increased age, a respiratory, systemic or enteric comorbidity as well as obesity were associated with a higher probability of mortality [166]. Obese individuals are predisposed to hypoventilation-associated pneumonia, pulmonary hypertension and cardiac stress, an increased risk of diabetes mellitus, cardiovascular disease and kidney disease [167]. The named comorbidities are considered to result in an increased vulnerability to pneumonia-associated organ failures [167]. Apart from this, the risk for type-2-diabetes patients to have a worse disease progression seems high according to a Chinese cohort study of 7, 337 patients with a confirmed SARS-CoV-2 infection among which 952 had type 2 diabetes [168]. A meta-analysis of multiple studies in China detected a 4-fold increase in mortality in patients with pre-existing COPD and confirmed SARS-CoV-2 infection [169]. Nevertheless, a German observational study with almost 1 million enrolled patients, of which 148, 557 had COPD and 818, 490 had asthma, did not deliver evidence for an increased mortality risk of COPD- or asthma-patients if they are treated with inhaled steroids [170]. With respect to studies concerning the impact of comorbidities it should generally be considered that the reporting of regarded cases might have concentrated on hospitalized and ICU patients rather than mild, outpatient infected people [171]. This raises case-fatality rates relatively in comparison to other countries.

As of July 22nd 2020, the most commonly detected comorbidities of COVID-19 patients in Italy were hypertension (66 % of the cases), type-2-diabetes (29.8 %), ischemic heart disease (27.6 %), atrial fibrillation (23.1 %), chronic renal failure (20.2 %), chronic obstructive pulmonary disease (17.1 %), active cancer during the last 5 years (16.3 %), heart failure (15.8 %) and obesity (10.7 %) [172].

German forensic medicals from Hamburg assumed that the number of unconfirmed SARS-CoV-2-induced deaths was lower than generally estimated because the dark figure of unconfirmed or asymptomatic or overcome infections was unresolved, whereas the dark figure of the proportion of fatal SARS-CoV-2 infections was enormously small [174]. In the northern German Federal State of Hamburg, all deaths of Hamburg citizens with ante- or post-mortem PCR-confirmed SARS-CoV-2 infection were examined post-mortem contrary to the initial recommendation of the RKI between the outbreak of the pandemic in Germany and May 15th 2020 [173]. The autopists provided a systematic overview of the first 80 consecutive full autopsies. They categorized the deceased people into definite COVID-19 deaths, probable COVID-19 deaths and possible COVID-19 deaths with an equal alternative cause of death. The deceased had an average age of 79.2 years, 34 were female and 46 male, and all of them except for 2 women had relevant comorbidities. The autopists found a COVID-19 pneumonia in 83 % of the cases and that a virus infection of the respiratory tracts and the lung was of central importance for fatal disease progressions [173]. Competing causes of death were considered in 11 % and clear causes of death not related to COVID-19 were found in 5 % of the cases [173]. Modifications due to bacterial superinfections, diverse comorbidities as well as multimorbidity in general usually completed the health profile of a person who died as a consequence of pneumonia. The typical symptom profile of an exclusive virus pneumonia was said to be exceedingly rare [174].

German pathologists from Kiel proved that 85 % of the deceased people with a confirmed SARS-CoV-2 infection, who were examined post-mortem at the institute of pathology of the university clinic Schleswig-Holstein, died due to COVID-19 and not simply with it. The team of pathologists performed more than 50 cases aged between 53 and 90 until the beginning of February 2021 [175].

It is obvious that the enumerated bias-triggering factors are associated with each other. If a CFR is used as an indicator of SARS-CoV-2 lethality, it is generally very likely to be overestimating, because a large fraction of infected people remains undetected as they are not included into the population of confirmed cases. Most often, merely symptomatic cases are tested and detected. However, a CFR may be an underestimating region-specific lethality rate concerning a certain time period in the way that the cumulative number of deaths might eventually keep growing as an increasing number of patients is hosted in intensive care units. The underestimation is caused by the time delay between the occurrence of the infection and the death in this case [163].

Another common error in CFR computation for a point in time t, that leads to misestimation, is the usage of the number of confirmed infected cases at time t and deaths of individuals with a confirmed infection at time t. The average time from case detection and confirmation to death is not incorporated then. If this time is indeed taken into account, it should firstly be considered that most cases are probably detected during their symptomatic period, which is after the incubation time, and secondly a bureaucratic delay is accountable for a lagged reporting to the responsible authorities. The average period between symptom development and death of a confirmed case was 11 days during the first wave of the corona virus pandemic in Germany according to the RKI, whereas this period was stated to last 16 or 18 days in other studies [10].

All of the described possible deviations of the foundations of CFR computation among countries complicate the assessability of the extent of mortality the infectious disease causes.

A possible way of achieving more realistic representations of lethality rates than by computing CFRs is using *infection-fatality rates* (IFRs). An IFR is obtained by dividing the number of confirmed as well as asymptomatic unconfirmed infected deaths by the number of known infected cases [176]. An analysis of age-specific IFRs found an exponential relationship between age and IFR for COVID-19 [177]. Whereas the estimated IFR for children and young adults was smaller than 1 %, it progressively increased at higher ages, reaching 15 % at age 85 [177].

To account for the deaths of asymptomatic unconfirmed infected people so-called *probable SARS-CoV-2 deaths* can be used. Whereas confirmed SARS-CoV-2 deaths can be defined as those those occurring in persons with a laboratory-confirmed

SARS-CoV-2 infection, probable SARS-CoV-2 deaths are, among others, those with COVID-19, SARS-CoV-2 or a similar term listed on the death certificate as an immediate, underlying, or contributing cause of death but did not have laboratory-confirmation of the infection [178]. If this kind of IFR is applied as an indicator of the actual lethality rate, it is inevitable to include probable infected cases into the divisor with the aid of a statistical estimation method in order to prevent overestimation. There are clinical findings proving that the population-based lethality rate of COVID-19 as measured by the IFR presumably lies between 0.5 to 1 %, which would mean that lethality is 5 to 10 times higher than for seasonal influenza [179].

Compared with seasonal influenza, COVID-19 was found to be associated with a higher risk of death, various secondary diseases, mechanical ventilator use, admission to intensive care, 3 additional days of hospital stay as well as 16.58 additional cases of death in hospitalized cases in an US-American cohort study between January 1st 2017 and December 31st 2019 that included 3.641 COVID-19 and 12, 676 influenza cases [180]. Among people admitted to hospital, compared with seasonal influenza, COVID-19 was associated with increased risk of extrapulmonary organ dysfunction, death, and increased health resource use [180].

Within the ICD-19-code-based hospital sentinel for severe acute respiratory diseases (ICOSARI), patients with a severe acute respiratory disease (SARI) are registered such that SARI-COVID-19 cases can be compared to SARI cases emerging from 5 different flu waves (SARI-GW cases). In a comparison of 1,426 SARI-COVID-19 cases of the year 2020 with 69,573 SARI-GW cases of the years 2015 to 2019, it was discovered that the fraction of serious disease progression, ventilated as well as deceased individuals was significantly higher in SARI-COVID-19 cases (21–22 %) than in SARI-GW cases (12–14 %) [181].

In a French nationwide retrospective cohort study using the French national administrative database, all patients hospitalized for COVID-19 between March 1st and April 30th 2020 (89, 530 patients) as well as all patients hospitalized for influenza between December 1st 2018 and February 28th 2019 (45, 819 patients) were included, where a classification as infected comprised a primary, related or associated diagnosis. It was observed that patients admitted to hospital with COVID-19 more frequently developed acute respiratory failure, pulmonary embolism, septic shock or haemorrhagic stroke than patients with influenza, but less frequently developed myocardial infarction or atrial fibrillation. Moreover, in-hospital mortality was almost 3 times higher in patients with COVID-19 (16.9 %) than seasonal influenza (5.8 %) [182]. Apart from this, a Swiss study among adult patients with COVID-19 or an influenza infection hospitalized in one of 14 regarded Swiss hospitals showed that the community-acquired COVID-19 was associated with worse outcomes compared to community-acquired influenza, as the hazards of in-hospital death and ICU admission were approximately 3 times higher [183]. An international study, which was described as significant, creative and sophisticated by the head of the institute of forensic medicine at the university clinic of Hamburg, compared autopsy and molecular pathological findings of COVID-19 and influenza patients that were all diagnosed with the acute respiratory syndrome (ARDS), and carried out postmortem examinations of deceased people without any respiratory infection. The influenza and COVID-19 cases showed diffuse changes in the alveoli of the lung epithelium with perivascular T-cell infiltration. Severe impairments of the endothelium with damaged cell membranes and 9 times higher risk for microthrombosis in the alveolar capillaries were associated with SARS-CoV-2 [184]. In summary, the results from the mentioned studies imply that SARS-CoV-2 has a higher potential for respiratory pathogenicity, leading to more respiratory complications and to higher mortality than influenza.

2.5.3   Excess Mortality

Apart from case-fatality rates excess mortality can be used as a means to estimate the seriousness and mortality effects of a lethal virus. It is a wider-ranging measure of the total impact of the pandemic on deaths than the confirmed death count alone [185]. Moreover, it refers to the number of deaths from all causes during a crisis above and beyond what would be expected to observe under "normal" conditions [185]. The number of excess deaths in a population over a certain time period is defined as the deviation in mortality from the expected non-crisis level [153]. Hence the baseline is the number of deaths expected when no particular process increases the mortality [153].

According to the European mortality monitoring activity EuroMOMO the prime mortality pattern in European countries is a Poisson distributed time series, which implies a trend and in some cases a sine cycle of a period of one year, and is modified by additional factors in winter and summer [153].

EuroMOMO makes use of Z-scores to quantify deviations from the basic mortality [186, p. 43]. Mortality baseline is modelled using a general linear Poisson corrected model for over-dispersion [153]. The following EuroMOMO graphic depicts the weekly average number of excess deaths in the EuroMOMO partner countries of all ages. The blue (grey, yellow, red) line represents the course of excess deaths numbers of the year 2020 (2019, 2018, 2017), and the dashed line is the zero function.

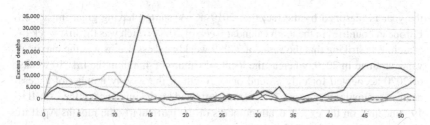

Figure 2.9 Weekly numbers of excess deaths in the data-providing EuroMOMO partner countries for all ages and the years 2020 (blue), 2019 (grey), 2018 (yellow) and 2017 (red). Source of data: [153]

When considering the course of the year in mortality statistics, the typical fluctuations during the influenza season from about mid-December to mid-April should be taken into consideration [187]. Additionally, it must be noted that the illustrated numbers of excess deaths comply with those of all EuroMOMO partner countries. The wave of influenza, which usually affects death numbers between mid-December and mid-April, was considered as finished in Germany in 2020 in the calendar week 12 according to the RKI [186, p. 44]. The impact of 2019-nCoV on mortality might be weakened by the fact that the influenza waves of the previous years took longer [186, p. 44].

It is conspicuous that the number of excess deaths in the EuroMOMO partner countries was lower in 2019 than 2018 and lower in 2019 than 2020 in the calendar weeks 1–10. This implies that the influenza wave at the turn of the year 2018/2019 (2017/2018) was more remarkable than the one of 2019/2020 (2018/2019), except for the calendar weeks 5 and 6, when there were around 1,000 more excess deaths in the year 2019 than 2018. During the first 10 calendar weeks of 2018, there were 2 local maxima given by 11,531 weekly excess deaths in week 1 and 11,353 in week 10. The ECDC stated that the influenza wave of 2017/18 had its activity peak in early January 2018 in south-western Europe and in mid-February 2018 in northern Europe, and excess mortality was mainly observed in people aged 65 years or older [188]. Figures were lower in almost the whole year 2017 than in 2019, with negative numbers in the first 16 calendar weeks.

The maximal number of weekly excess deaths attained in the complete year 2019 is 7, 299, that was reached in the calendar week 6. The maximal value achieved in the first 10 calendar weeks of 2020 is 4, 898 (calendar week 2), which is only 2 thirds of the local maximum of the first 10 weeks in 2019. The low level of death numbers in European countries until March 2020 in comparison with the rest of the year is reasoned by the fact that 2019-nCoV started having great impact on European countries in the tenth calendar week of 2020, which was the first week of March. It is striking that the value of 7, 299 weekly excess deaths was the maximal value attained in 2019, whereas the local maximum of 4, 898 in the first 10 weeks of 2020 accounted for only around one seventh of the weekly number of excess deaths maximally reached in 2020. This is effected by the influence of the COVID-19 pandemic on the excess death numbers of the particularly the months April to May and October to December 2020.

The course and local maximum of the weekly number of excess deaths in the first 10 weeks of the years 2018 and 2019 was a lot higher than the highest local maximum of the rest of the respective year (week 31 and difference of 7,688 in 2018, week 30 and difference of 3,349 in 2019). The same holds for the local maximum of week 52 of 2017 compared to the highest local maximum of the rest of the same year (week 25 and difference of 6,074). This can also be reasoned by the typical annual influenza-effected fluctuations.

In the calendar week 10 of 2019, there were 2, 171 and in week 11 of the same year there were 1, 262 excess deaths, whereas only 1, 220 excess deaths were registered in week 10 of 2020 but 3, 392 in week 11 of the same year. This approximately equals a 0.58-fold decrease in 2019 and 2.78-fold increase in 2020 between the calendar weeks 10 and 11.

The extreme increase in excess deaths between the calendar weeks 10 and 14 of 2020 cannot be explained by normally occurring fluctuations. A value of 35, 335 was attained in week 14 in 2020, whereafter numbers decreased until they reached and remained on a level that was similar to the ones of the previous years during the calendar weeks 20 to 40, fluctuating between values of −614 (week 25) and 5, 608 (week 33). Another deviation, not explainable by common fluctuations according to the values of the 3 previous years, can be observed in the calendar weeks 41 until 52 of 2020, where a local maximum of 15, 876 was reached in the calendar week 46. This value is higher than the highest local maximum of 3 previous years (11, 531, calendar week 1 in 2018). The worldwide spread of the novel corona virus is a logical explanation for the abnormal sharp increases in excess death numbers in spring and autumn 2020.

A question arising from Figure 2.9 is why the number of excess deaths of 2020 was exceedingly larger in spring than in autumn/winter. The local maximum of excess deaths was lower in autumn/winter than in spring 2020 by around 20, 000, although Europe experienced higher numbers of new infections as well as deaths in autumn than spring 2020. According to the online data source Ourworldindata.org, the maximal number of daily newly confirmed deaths in the first wave of the pandemic was lower by around 450 in the European Union and 1,530 in Europe [90].

The combination of less excess deaths but more confirmed COVID-19 deaths in European countries in autumn/winter compared to spring 2020 indicates a general annually higher level of deaths in autumn/winter than in spring in Europe. Another possible explanation for less excess deaths is the better preparation for the second wave of the pandemic compared to the first wave. For instance, the European Union planned to test more efficiently using rapid antigen detection, spread the newest information on the pandemic quicker across Europe, trace infections via app usage and provide tax-free medical equipment and apply vaccination strategies as soon as possible before the outbreak of the second wave [189]. Substantially ICUs play a major role in the management and treatment of critically ill individuals.

The German RKI determines the daily supply capacities of all German hospitals with intensive care and numbers of cases in intensive care beds units [190].

Furthermore, it was shown that the medication dexamethasone achieved success in the treatment of severely affected patients already in summer 2020. This might explain a reduced mortality risk in intensive care patients in the second wave compared to the first wave, which may be one of several causes of less excess deaths despite of higher numbers of patients in ICUs. The treatment with dexamethasone was the first medication lowering mortality among ICU patients in need of medical respiration [191].

A controlled, open-label trial study including 6, 425 patients, of which some were randomly assigned patients to receive oral or intravenous dexamethasone, verified that dexamethasone reduced 4-week mortality among those receiving invasive mechanical ventilation or oxygen at randomization, but not among patients not receiving respiratory support [192, 193].

It may make sense to use the courses of excess death numbers as a basis of the evaluation of the impact of COVID-19 on populations instead of CFRs. The reason is that excess death number computation takes the average number of deaths in a population within a certain time period as a basis, whereas CFR computation is based on the numbers of confirmed infected and confirmed infected deceased individuals. Hence excess mortality is computed independently of the incidence of the specific disease in order that it is a comprehensive measure of the total impact of the pandemic on deaths. Additionally, the described uncertainties of CFR computation could have effected misestimations between the country-specific CFRs of spring and autumn 2020, substantially because populations had different reactions to the second than the first wave of the pandemic.

However, it must be taken into account that EuroMOMO uses a complex, scientific approach of linear, generalized Poisson regression models and analyses various European countries at once such that means of standardization have to be found

[194]. This involves restrictions concerning the death numbers of the single countries.

The German Federal Statistical Office developed a method of measuring excess mortality that differs from the EuroMOMO approach. It uses a simple descriptive comparison of the current mortality rate with the average mortality rate of the previous years [186, p. 43].

Figure 2.10 shows the computed weekly deaths in Germany, where the blue line represents the average of the years 2016 to 2019, the bright blue area symbolizes the bandwidth of the years 2016 to 2019, and the red line illustrates the course of death numbers in 2020. The red line on the bottom represents the number of confirmed COVID-19 deaths.

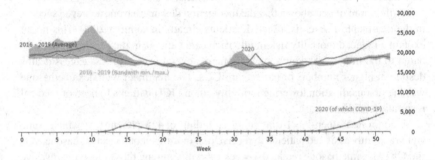

Figure 2.10 Weekly numbers of deaths in Germany for all ages and the years 2020 (red) and 2016 to 2019 (blue), bandwidth of deaths in Germany between 2016 and 2019 (bright blue), and weekly numbers of German COVID-19 deaths in 2020 (dotted red). Source of data: [194]

Figure 2.10 conveys strongly increased numbers of death cases as well as COVID-19 deaths between the calendar weeks 13 and 18 in 2020. The maximal value of 20, 646 in spring was reached in the calendar week 14, where the number of excess deaths in the EuroMOMO diagram is highest. A sharp increase between the calendar weeks 12 and 20 is also observable in the EuroMOMO graph. The deviation from the average weekly number of deaths of the previous 4 years was approximately 14 % in the calendar week 15 in 2020. This is substantially effected by the age-specific mortality rates of people over 80 years [186, p. 45].

A report of the Federal Statistical Office stated that comparisons between 2020 and the years 2016 to 2019 were expedient from the calendar week 12 because of the influenza wave that lasted until the calendar week 12 in 2020. In particular, the influenza wave seemed to have resulted in a maximal number of deaths of 23,640 in the calendar week 5 in 2017, 26,777 in the calendar week 10 in 2018, and 20,737 in the calendar week 9 in 2019, but had comparatively little influence on the death numbers in Germany in 2016 and 2020 [186].

In accordance with the statements of the Federal Statistical Office, the noticeably high death numbers and simultaneous high incidence in Germany between the calendar weeks 12 and 20 of 2020 lead to the view that causes of death other than COVID-19 can only have played a subordinate role for this development.

In the above diagram, it is conspicuous that there is a clear local maximum in the calendar week 33 in August 2020, which is 3, 356 more deaths than the average of the previous years. This coincides with the depiction of the EuroMOMO diagram including all EuroMOMO partner countries, in which a local maximum of 5, 608 excess deaths is shown for the same week. From the calendar week 41 (first week of October), the weekly death numbers rise steadily, which is a similar development as in the EuroMOMO graph. The number of individuals who died and had laboratory confirmed COVID-19 disease increases at the same time. In the first half of October, the total number of weekly deaths declines to the average of the previous 4 years. When the number of COVID-19 deaths starts growing from week 41, there is also an increase in total deaths, in excess of the average. In October 2020, they are by 5 % and in November by 12 % above the average of the previous years. A value of 23, 550 newly confirmed weekly deaths is attained in week 51 in 2020, which is around a quarter of the average of the years 2016 to 2019 and an all-time high.

The maximum number of excess deaths in spring in the EuroMOMO diagram is larger by around 20, 000 than the local maximum of autumn/winter, whereas the maximum number of excess deaths in spring of the diagram of the Federal Statistical Office is smaller by less than 3, 000 deaths than the local maximum in autumn/winter. Here the country-specification of Figure 2.10 and also the different computation methods seemingly become obvious.

In addition to Figures 2.9 and 2.10, a German study examining age-specific excess mortality due to COVID-19 demonstrated that the SARS-CoV-2 pandemic produced excess mortality in Germany during its first wave from the calendar week 10 to the calendar week 23 in 2020 [150].

This agrees with the messages deducible from Figures 2.9 and 2.10. The study also discovered that the largest increase in deaths between 2016–2019 and 2020 occurred among people aged 80–89 and over 90 [150].

The SIR Model in Epidemic Modelling

3

SARS-CoV-2 is a pandemic, which is defined as a transnational human infectious disease that spreads across different continents. Mathematical models of infectious diseases can be constructed with the aid of compartment models, that divide the underlying population into different groups, which are called compartments. Every individual in the same compartment has the same characteristics.

In those models, individuals can be in a finite number of discrete states, of which some are simply labels that specify the various traits of individuals, some are changing with time, such as age class, some are fixed, such as sex or species, and others indicate the progress of an infection [195, p. 873].

The transitions of individuals between different compartments can be expressed through systems of ODEs. This chapter gives an introduction to the so-called SIR model and simple extensions in order to be able to set up a SARS-CoV-2-fitted model in Chapters 4 and 5. The correspondent system of ODEs is integrated in $MATLAB$ in Chapter 6.

In Section 3.1, an explanation of the so-called $SIR\ model$ is given, involving assumptions of this model, the model-specific maximally possible number of infections and incidence rates. In Section 3.2, simple extensions to the SIR model are revealed, which are characterized by population compartments added to the model.

All of the compartment models can be applied to networks, which are systems whose structure can be mathematically modelled with the aid of graphs, that are ordered pairs $G = (\bar{V}, \bar{E})$ with $\bar{V}$ a set of nodes (also called vertices) and $\bar{E}$ a set of edges. In this case the nodes can represents different geographic regions in order to for example model the transmission through travelling, see [196, 197]. This possibility is detailed in the conclusion of this thesis.

© The Author(s), under exclusive license to Springer Fachmedien Wiesbaden GmbH, part of Springer Nature 2021
S. M. Treibert, *Mathematical Modelling and Nonstandard Schemes for the Corona Virus Pandemic*, BestMasters, https://doi.org/10.1007/978-3-658-35932-4_3

3.1   The Basic SIR Model

Compartment models originate from the significant work of Kermack and McKendrick, who published a deterministic epidemic model including susceptible, infected, and removed individuals in 1927 [198]. This model still serves as the basic form of the SIR model today.

A deterministic model can be described by a set of ODEs in a single system, and assumes a homogeneous population, which can be regarded as a population in which individuals mix uniformly and randomly with each other. For a homogeneous population, it can either be assumed that all susceptibles of the same susceptible compartment have comparable contact patterns, or, as in this thesis, have an equal probability of contacting every other individual in the population at every point in time, and people exclusively differ in their disease state [199]. Moreover, the output of a deterministic model is fully determined by the parameter values and the initial conditions. In the sequel, the word "*system*" (except for the "system of ODEs") describes all of the individuals of a regarded population, which can be the inhabitants of a regarded country, as well as their transitions between certain compartments. The word "*population*" encloses the number of inhabitants who activate the dynamics in the system, regarding a specific country or certain regions.

3.1.1   Definition of the Basic SIR Model

Subsection 3.1.1 defines the SIR model by outlining its assumptions. The term *Basic SIR model* is used throughout this thesis in order to distinguish the typical SIR model from other (enhanced) compartment models introduced in this thesis. In Subsection 3.1.2, the maximal size of infected individuals in the Basic SIR model is computed, and in Subsection 3.1.3 different forms of so-called *incidence rates* are explained.

Assumptions of the Basic SIR Model
The simplest form of a compartment model is the SIR model, which consists of the three compartments (or: classes) of susceptible (S), infected (I) and recovered (R) individuals. The change of the number of compartment members over time is expressed by expressing the classes by time-dependent functions $S(t)$, $I(t)$ and $R(t)$. Susceptibles have not contracted but can contract the disease. Infected individuals have already contracted the infection. In the Basic SIR model, they are also able to infect susceptible individuals. Hence, they are assumed to be infectious and possibly

but not necessarily symptomatic. Recovered individuals have overcome the illness and are not sick with it anymore. Four basic assumptions (A1)-(A4) are made in this specific model [200, pp. 10+13]:

(A1) No births and no deaths are incorporated into the system.
(A2) The population is closed, so nobody enters any compartment from outside or leaves the system.
       The total population size remains constant.
(A3) Recovered people are completely immune such that they cannot contract the disease ever again.
(A4) Infected individuals are also infectious.

These assumptions can be explained as follows.

(A1) In many cases the dynamics of an infectious disease are much quicker than the system inflow, which for instance is a consequence of birth or mortality dynamics. Therefore, birth and death rates are often omitted in simple forms of compartment models. Disease-induced death does not play any relevant role for many epidemics such that it can be omitted as well. In the case of conceivably lethal diseases spreading quickly across large parts of or even the whole world population and leading to a considerable relative amount of deaths, disease-induced deaths can nevertheless be interesting or even important to include. An example is the SARS-CoV-2 pandemic, which is modelled in Chapters 4 and 5.

(A2) The size of the population at a point in time t is given by $N(t)$. The fulfilment of the equation

$$N(t) = S(t) + I(t) + R(t) \quad \text{with } N : [0, T] \to \mathbb{N},$$

means that the number of individuals in the system is the sum of the compartment sizes at each regarded time instant $t \in [0, T]$. The system has to be equipped with initial conditions $S(0)$, $I(0)$, $R(0)$ in order to be well-defined [200, p. 11]. The constancy of the population size $N(t)$ mentioned in (A2) is given if the derivative of $N(t)$ is zero:

$$N'(t) = S'(t) + I'(t) + R'(t) = 0 \quad \forall t \in [0, T] \Rightarrow N(t) = N = const.$$

(A3) If a permanent cure was not assumed, there would be different options to model the transfer of recovered or infectious people back to the susceptible state. Options to do this are detailed in Subsection 4.2.9.

(A4) In the case that infected individuals are not automatically infectious, for instance if a certain incubation time of the disease is given, the compartment I has to be split into (at least) two compartments, of which one contains infected but not infectious individuals and the other one comprises infectious people. To regard a modelled disease as infective, the infectiousness of at least one compartment is required. It is possible to include more than two infected or/and infectious compartments if different stages of disease progression exists. For example, there may be symptomatic and asymptomatic or discovered and undiscovered infected cases in the population. With respect to the novel corona virus, several stages can be distinguished. In Chapter 4, it is explained which compartments are used to model the dynamics of SARS-CoV-2 and COVID-19 in this thesis.

Formulation of the Basic SIR Model

Let $\gamma = \bar{\gamma} \cdot N(t)$ represent the number of contacts an individuals makes per unit of time t. Here, $\bar{\gamma}$ is the *per capita contact rate*. Let β represent the probability of a contact leading to the transmission of the regarded infection. It is called the *transmission risk* here. If the per capita contact rate $\bar{\gamma}$ is multiplied by β, the rate $\tilde{\beta}$ is obtained, which is the rate at which susceptible individuals are infected per unit of time:

$$\tilde{\beta} = \beta \cdot \bar{\gamma} \ .$$

Since the number of susceptibles who get infected per infectious individual per unit ot time t is

$$\beta \cdot \left(\bar{\gamma} \cdot N(t) \right) \cdot \frac{S(t)}{N(t)} = \tilde{\beta} \cdot S(t) \ ,$$

where $\frac{S(t)}{N(t)}$ is the fraction of susceptibles within the population, the term

$$\tilde{\beta} \cdot I(t) \cdot S(t)$$

represents the number of susceptibles who get infected at time t, where $\tilde{\beta} \cdot I(t)$ describes the so-called force of infection of the respective disease [200, p. 10]. The proportion of infective individuals who recover per time instant is given by the product $\omega \cdot I(t)$, where ω symbolizes the recovery rate. The system of ODEs arising from the mentioned compartments and assumptions is given in the Equations (3.1) to (3.3).

$$\frac{dS(t)}{dt} = -\tilde{\beta} \cdot S(t) \cdot I(t) \ , \tag{3.1}$$

$$\frac{dI(t)}{dt} = \tilde{\beta} \cdot S(t) \cdot I(t) - \omega \cdot I(t) \ , \tag{3.2}$$

$$\frac{dR(t)}{dt} = \omega \cdot I(t) \ . \tag{3.3}$$

The resulting model is nonlinear due to the product of the functions $I(t)$ and $S(t)$ in the Equations (3.1) and (3.2), which symbolizes a nonlinear (quadratic) dependence. It is furthermore dynamic because time-dependent changes in system states are included. All compartment models regarded in the Chapters 3 and 4 are deterministic since all sets of variable states are uniquely determined by the model parameters and the initial states of the variables [200, p. 8]. Stochastic compartment models are an alternative to deterministic models, and typically used types of stochastic model formulations are discrete time Markov chain and continuous time Markov chain models. Markov chain epidemic models are introduced in Chapter 6 of this thesis.

The incidence of the infectious disease is proportional to the product of $S(t)$ and $I(t)$ with regard to the model in (3.1) to (3.3). It is said that in this case the incidence is bilinear and given by the law of mass action [200, p. 37]. The term *incidence* itself as well as other forms of incidence are explained in Subsection 3.1.3.

If there was no inflow into the compartment I, the following differential equation for the dynamics of the class I would be obtained:

$$I'(t) = -\omega \cdot I(t) \text{ with } I(0) = I_0 \ .$$

This simple ordinary differential equation is solved by

$$I(t) = I_0 \cdot e^{-\omega \cdot t}$$

such that $\frac{I(t)}{I_0} = e^{-\omega \cdot t}$ is the proportion of people who are still infectious at time instant t and

$$F(t) := 1 - e^{-\omega \cdot t}, \qquad t \geq 0$$

is the proportion of people who have already recovered until time t [200, p. 16]. It is a probability distribution of the random variable t that represents the time to leaving the infectious class.

A *probability density function* (PDF) is defined as a function $f_X : \mathbb{R}^{n_x} \to \mathbb{R}, x \longmapsto f_X(x)$ with X a real-valued random variable and $x \in \mathbb{R}^{n_x}$. With reference to the relation between the PDF and the probability of X lying in the interval $[a, b]$ it holds that

$$\mathbb{P}(X \in [a, b]) = \int_a^b f_X(x) dx \ . \tag{3.4}$$

Equation (3.4) can be reformulated to obtain the definition of f_X:

$$f_X(x) = \lim_{\Delta x \to 0} \frac{P(X \in [x, x + \Delta x])}{\Delta x}.$$

It follows that the PDF of $F(t)$ is denoted by

$$f(t) = \frac{dF(t)}{dt} = \omega \cdot e^{-\omega \cdot t}, \quad t \geq 0$$

and the expected value of the time to exiting the class I (that is X_{ex}) is given by

$$\mathbb{E}[X_{ex}] := \int_0^\infty \left(1 - F(t)\right)dt - \int_{-\infty}^0 F(t)dt = \int_{-\infty}^\infty t \cdot f(t)dt = \int_{-\infty}^\infty t \cdot \omega \cdot e^{-\omega t}dt = \frac{1}{\omega}.$$
$$(3.5)$$

Since $\mathbb{E}[X_{ex}]$ can be interpreted as the mean duration of the infectious period, it has been proven that the mean time spent in the infected compartment is $\frac{1}{\omega}$ [200, p. 16].

3.1.2   Maximum number of infections in the Basic SIR Model

A significant question to be answered is what the maximal number of infected individuals over the course of time is and when it is reached. It is obvious that in the Basic SIR model prevalence starts to increase if

$$I'(0) = \tilde{\beta} \cdot I(0) \cdot S(0) - \omega \cdot I(0) = \left(\tilde{\beta} \cdot S(0) - \omega\right) \cdot I(0) > 0.$$

According to the above inequality and due to the fact that $I(0) \geq 0$ as well as the assumption $I(0) \neq 0$, a condition sufficient for an initial increase in prevalence is given by the following inequality:

$$\tilde{\beta} \cdot S(0) - \omega > 0 \quad \Leftrightarrow \quad \frac{\tilde{\beta} \cdot S(0)}{\omega} > 1.$$

The Equations (3.1) and (3.2) of the given system of ODEs do not depend on R, wherefore Equation (3.3) is omitted in the sequel. Dividing (3.1) by (3.2) we obtain an equation for I' independent of I [200, p. 14].

$$\frac{I'(t)}{S'(t)} = \frac{dI}{dS} = \frac{\tilde{\beta} \cdot S \cdot I - \omega \cdot I}{-\tilde{\beta} \cdot S \cdot I} = -1 + \frac{\omega}{\tilde{\beta} \cdot S} \quad \Leftrightarrow \quad dI = \left(-1 + \frac{\omega}{\tilde{\beta} \cdot S}\right) \cdot dS .$$
$$(3.6)$$

Integrating Equation (3.6) with respect to $S \in [0, \infty)$ yields

$$I_\infty - I_0 = -(S_\infty - S_0) + \frac{\omega}{\tilde{\beta}} \cdot \ln\left(\frac{S_\infty}{S_0}\right) \tag{3.7}$$

with $S_0 = S(0)$, $I_0 = I(0)$ and $S_\infty = \lim_{t \to \infty} S(t)$ the so-called *"final size of the epidemic"*.

If (3.6) is integrated up to only a finite value of S, we obtain

$$I(S) = S_0 + I_0 - S + \frac{\omega}{\tilde{\beta}} \cdot \ln(S) .$$

Moreover, we divide Equation (3.1) by (3.3):

$$\frac{S'(t)}{R'(t)} = \frac{dS}{dR} = -\frac{\tilde{\beta} \cdot S}{\omega} . \tag{3.8}$$

The integration of Equation (3.8) yields

$$S(R) = S_0 \cdot e^{-\frac{\tilde{\beta}}{\omega} \cdot R} \geq S_0 \cdot e^{-\frac{\tilde{\beta}}{\omega} \cdot N} > 0 . \tag{3.9}$$

It can be concluded that S_∞ is always positive such that there will always be susceptible people in the population. Integrating Equation (3.1) we obtain

$$S_0 - S_\infty = \tilde{\beta} \cdot \int_0^\infty I(t) \cdot S(t) dt$$

and if $I(t)$ is integrable on $[0, \infty)$, the equation

$$S_0 - S_\infty \geq \tilde{\beta} \cdot S_\infty \cdot \int_0^\infty I(t) dt \tag{3.10}$$

is yielded. Using Equation (3.10) it is obtained from Equation (3.7) that

$$I_0 = S_\infty - S_0 + \frac{\omega}{\tilde{\beta}} \cdot \ln\left(\frac{S_0}{S_\infty}\right) \quad \Rightarrow \quad \frac{\tilde{\beta}}{\omega} = \frac{\ln\left(\frac{S_0}{S_\infty}\right)}{S_0 + I_0 - S_\infty} .$$

The maximal number of infected individuals is reached if $S = \frac{\omega}{\tilde{\beta}}$. Using this fact, the maximal size of the infected compartment I_{max} occurring in the epidemic (or pandemic) [200, p. 15] is given by

$$I_{max} = -\frac{\omega}{\tilde{\beta}} + \frac{\omega}{\tilde{\beta}} \cdot \ln\left(\frac{\omega}{\tilde{\beta}}\right) + S_0 + I_0 - \frac{\omega}{\tilde{\beta}} \cdot \ln(S_0) \, .$$

3.1.3 Incidence Rates

In Subsection 3.1.1 a bilinear incidence was used for the Basic SIR model. The general definition of incidence is introduced at first, before different forms of incidence rates are distinguished.

Incidence is the number of new infections occurring within a defined time span [147, p. 33], i.e. is represented by the number of individuals transiting from the susceptible to the infected compartment per unit of time in the SIR model. An incidence rate is a measure of the frequency with which a disease occurs over a specified time period. It can be defined by the ratio of the number of new infections to the average of the population size in the observed time period. It is significant to notice that incidence is often determined from the number of clinical cases but not subclinical i.e. asymptomatic cases. This results in an underestimation of diseased cases.

The application of different incidence rates can change behaviours of the underlying system [201]. Different kinds of incidence rate are stated in (I1) to (I4).

(I1) Mass action incidence: $\tilde{\beta} \cdot S \cdot I$,

(I2) Standard incidence: $\tilde{\beta} \cdot \dfrac{S}{N} \cdot I$,

(I3) Incidence saturating in the size of infected: $\tilde{\beta} \cdot \dfrac{S \cdot I}{1 + \sigma_1 \cdot S}$ with $\sigma_1 > 0$,

(I4) Incidence saturating in the size of susceptibles: $\tilde{\beta} \cdot \dfrac{S \cdot I}{1 + \sigma_2 \cdot I}$ with $\sigma_2 > 0$.

Mass action incidence (I1) is implied by the law of mass action, "analogously to terms from chemical kinetic models, whereby chemicals react by bumping randomly into each other" [200, p. 37]. The standard incidence (I2) differs from mass action incidence (I1) in the way that $\tilde{\beta} \cdot S \cdot I$ is normalized by the total population size N. If a constant population size $N(t) = N$ for all t is assumed, the mass action incidence and standard incidence coincide [200, p. 37]. However, they differ if a non-constant total population size is assumed.

The bilinear incidence rate can be normalized by the term $N(t) - D(t)$ instead of $N(t)$ if a deceased compartment is included in the model.

In the saturating incidence rates $(I3)$ and $(I4)$, inhibition effects decrease the proportion of susceptibles transiting to the infected class per unit of time contingent on the parameter σ_1 or σ_2, respectively [201]. These effects are incorporated with the aid of the term $\frac{1}{1+\sigma_1 \cdot S}$ or $\frac{1}{1+\sigma_2 \cdot I}$, respectively. In $(I3)$, inhibition results from behavioural changes of the susceptibles accounted for by an increasing susceptible population [201]. In $(I4)$ it is caused by a crowding effect of the infected individuals. This means that the fraction of susceptibles moving to the class I is negatively affected by the size of the population of already infected individuals [201]. Since $(I3)$ and $(I4)$ involve changing behaviours of S or I, respectively, they may seem more reasonable than $(I1)$.

If the limits of $(I3)$ and $(I4)$ are regarded as S or I tend to infinity, respectively, the following incidence limits are obtained:

$$\lim_{S \to \infty} \tilde{\beta} \frac{I \cdot S}{1 + \sigma_1 \cdot S} = \frac{\tilde{\beta} \cdot I}{\sigma_1},$$

$$\lim_{I \to \infty} \tilde{\beta} \frac{S \cdot I}{1 + \sigma_2 \cdot I} = \frac{\tilde{\beta} \cdot S}{\sigma_2}.$$

It follows that the limit of the incidence rate is independent of S for $(I3)$ or I for $(I4)$. Consequently, the number of effective contacts between infective and susceptible individuals saturates at high infective levels. Additionally, a combination of the incidence rates $(I3)$ and $(I4)$ in the form $\tilde{\beta} \cdot \frac{S \cdot I}{1+\sigma_1 \cdot S+\sigma_2 \cdot I}$ with $\sigma_1, \sigma_2 > 0$ is possible as well.

The mass action incidence is often used in diseases for which disease-relevant contact increases with an increase in the population size. For instance, in influenza and SARS, contacts increase as the population size and density increase [200, p. 38].

3.2   Simple Enhancements to the Basic SIR Model

For different infectious diseases various other compartments can be added to the model. For simplicity, a constant total population is assumed for all of the extended models in this section.

3.2.1 $SIRD$ Model

In the $SIRD$ model a compartment D of deceased individuals is added. $D(t)$ comprises all individuals who have died up to the time t due to their infection. Secondary disease can be included as reasons for transition to the compartment D. In this case, it can be said that infection-related reasons lead to a transition to the deceased compartment. Individuals can only enter D from I in the $SIRD$ model. An application of this model has the advantage that a constant total population size and deceased individuals can be assumed simultaneously because the deceased ones are transferred into a separate compartment. In the Basic SIR model infection-related deaths are omitted or neglected or can be regarded as included into the rate μ if a natural death rate μ is involved.

3.2.2 $SEIR$ Model

A commonly occurring supplementary compartment is E, which represents the class of exposed individuals. Here, individuals have been contaminated but are not yet able to infect others. They are not contagious. In this case the compartment I does not comprise individuals who are not infectious. A new parameter χ represents the rate at which individuals move from E to I, so get infectious per unit of time. All individuals in the compartment E are passing the latent period, which has to be separated from the incubation period. Enhancing the above SIR model to a $SEIR$ model, the outcome is the following:

$$\frac{dS(t)}{dt} = -\tilde{\beta} \cdot S(t) \cdot I(t) \,,$$
$$\frac{dE(t)}{dt} = \tilde{\beta} \cdot S(t) \cdot I(t) - \chi \cdot E(t) \,,$$
$$\frac{dI(t)}{dt} = \chi \cdot E(t) - \omega \cdot I(t) \,,$$
$$\frac{dR(t)}{dt} = \omega \cdot I(t) \,.$$

Susceptibles are infected as a result of a contact with at least one infectious individual. Infected individuals do not directly move from S to I but firstly enter the latently infected state. As mentioned above, birth and death rates are often omitted and/or regarded as equal in compartment models, such that the total population has a constant size N. This is especially true for low lethality diseases. For the SEIR model, this yields:

$$S(t) + E(t) + I(t) + R(t) = N = const. \qquad \forall t \in \mathbb{N}.$$

Every model with a compartment I can be expanded by a class E. Regarding the SARS-CoV-2 pandemic considered in the further sections of this thesis, the latent period and the incubation period are included in the model with the aid of additional compartments. This is stated in more detail in Chapter 4.

3.2.3 $SEIS$ Model

In the $S(E)IS$ model, an infection does not leave any immunity to recovered individuals who subsequently transit back to the compartment S. It is important to notice that initially susceptible and recovered susceptible individuals are both comprised by the compartment S here. Let the parameter ϖ describe the rate at which people transit from the recovered to the susceptible state per unit of time t. The $SEIS$ model with a constant total population size can be established as in the Equations (3.11) to (3.13).

$$\frac{dS(t)}{dt} = -\tilde{\beta} \cdot S(t) \cdot I(t) + \varpi \cdot I(t) , \qquad (3.11)$$

$$\frac{dE(t)}{dt} = \tilde{\beta} \cdot S(t) \cdot I(t) - \chi \cdot E(t) , \qquad (3.12)$$

$$\frac{dI(t)}{dt} = \chi \cdot E(t) - \omega \cdot I(t) - \varpi \cdot I(t) . \qquad (3.13)$$

This model is reasonable if the regarded disease does not confer any long-lasting immunity to recovered individuals, and recovered people have the same risk of getting infected as susceptibles who have not recovered. A non-lasting, declining or vanishing force of protection given by recovery is often referred to as *"waning"*.

3.2.4 Stages Related to Disease Progression and Control Strategies

The SIR model can be enlarged for several reasons differing from the ones introduced in Subsection 3.2.3. One reason is the fact that a transmission risks may originate from multiple infected compartments instead of one.

Firstly, so-called carriers can be added to a compartment model. The carrier compartment comprises individuals who are not sick themselves and do not show any symptoms but are able to transmit the pathogen. The carrier compartment is often referred to as the asymptomatic compartment of model. The carrier state A can be added between compartments S and I or E and I. A transition from A to R should be added if the respective infection can progress without any symptoms.

An infection without symptoms may occur as an alternative to symptomatic infectiousness. In this case exposed individuals do not transit from A to I, but move to the symptomatic infectious compartment I with probability rate p, and to the asymptomatic infectious compartment A with the probability rate $(1 - p)$. Here p lies in the interval $[0, 1]$. Examples for epidemics with necessary modelled asymptomatic compartments are malaria or HIV [200, p. 94].

Instead, the carrier compartment can be inserted between the compartments I and R in order to gather those individuals who are still infectious but do not show symptoms anymore before moving to the recovered class. Another possible interpretation of the carrier compartment is that some people never completely recover and continue to carry the infection, i.e. the class of carriers replaced the recovered class.

Another possibility of different disease progression is passive immunity. It describes latency before entering the susceptible compartment. For this reason it is added to the model by placing individuals in a compartment M, from which they progress to the susceptible stage S at a certain rate. "Passive immunity is the transfer of active immunity in the form of antibodies from one individual to another. Passive immunity can occur naturally, when maternal antibodies are transferred to the foetus through the placenta or in the milk during breastfeeding. Passive immunity can also be induced artificially, when high levels of antibodies specific for a pathogen or toxin are transferred to nonimmune individuals." [200, p. 94].

Apart from enhancements based on disease progression the S(E)IR and S(E)IS models can be expanded by the inclusion of disease control strategies. At first, a compartment of quarantine is introduced here. Quarantine can be imposed on susceptible citizen for a defined time period as a measure to limit the number of disease transmissions and also protect the people in quarantine from contagion. It is significant to consider if the respective disease spreads quickly across the whole regarded population and responsible politicians or health authorities actually apply quarantine measures.

An example for an epidemic requiring quarantine measures is COVID-19. As seen during the course of the spread of 2019-nCoV, governments can retain whole population groups such as repatriates in quarantine. Moreover, self-quarantine can be considered.

Quarantine has to be distinguished from isolation, which is here defined as a confinement of an infectious individual, which restricts contacts of the isolated individual with other individuals in the population. Both quarantine and isolation are included in the SARS-CoV-2-fitted model in Chapter 4.

Moreover, a treatment stage can be included. Individuals either progress to the recovered or treatment class when leaving the infectious compartment [200, p. 95]. This is reasoned by the fact that incomplete or not fully successful treatment of patients can be responsible of relapses, which cause individuals return from the recovered/treated or infectious to the exposed compartment E. A certain portion of the people in the treatment compartment can transit back to E at a relapse rate.

Beyond that, a transition of exposed or infectious individuals to the treatment class can be added then for the purpose of modelling treatment starting before infectiousness, which means before individuals enter the compartment I.

The last control strategy to be mentioned here is vaccination. Vaccination may provide complete or partial immunity to the infectious disease [200, p. 96]. Details concerning the incorporation of vaccine into the model are stated in Subsection 4.2.11, since vaccination is a relevant feature in the SARS-CoV-2-fitted model.

The SARS-CoV-2- tted SEIR Model

4

This chapter presents a dynamical enhanced compartment model, that is independently created on the basis of the population groups and dynamics observable in populations that are attacked by 2019-nCoV. It has been designed to meet the requirements and adapt to the specificities of SARS-CoV-2 and COVID-19, and is named the *SARS-CoV-2-fitted model* throughout this thesis.

The main target of this chapter is to create a compartment model framework reflecting various observed SARS-CoV-2-based population dynamics in a realistic way. This framework is composed of the concepts of all sections of Chapter 4, including the basic assumptions of the SARS-CoV-2-fitted model (Section 4.1), the characteristics of its compartments (Section 4.2), and the transition and transmission rates included in the dynamical model (Section 4.3).

Section 4.1 gives an overview of the differences between the assumptions made for the Basic SIR model in Chapter 3 and the SARS-CoV-2-fitted model. All of the 13 compartments of the SARS-CoV-2-fitted model are explained in Section 4.2. Section 4.3 introduces mathematically defined rates concerning the transitions between different compartments, which are composed of a transmission risk, a contact rate, an isolation rate and a quarantine rate. These rates can be modelled by time-dependent functions in order to better adapt the SARS-CoV-2-fitted model to the real dynamics of the pandemic. The mathematical formulation of the transmission rates used in the implemented models is reasoned in regard of the progression of COVID-19 in the world in the year 2020.

S. M. Treibert, *Mathematical Modelling and Nonstandard Schemes for the Corona Virus Pandemic*, BestMasters, https://doi.org/10.1007/978-3-658-35932-4_4

4.1   Differences to the Basic SIR Model and Model Assumptions

The SARS-CoV-2-fitted $SEIR$ model enhances the Basic SIR model by several compartments in order to better depict the infection-based states that individuals are assumed to traverse under the circumstance of the SARS-CoV-2 pandemic. In this section, the differences between the Basic SIR model presented in Chapter 3 and the developed SARS-CoV-2-fitted model are explained. Apart from added compartments the model assumptions of the newly introduced model differ from the assumptions of the Basic SIR model described in Subsection 3.1.1.

The following basic assumptions (B1)-(B4) are established for the model for the purpose of distinguishing it from the assumptions (A1)-(A4) of Section 3.1.

(B1) A natural death rate μ is incorporated into the model. It is the rate at which individuals die per unit of time for reasons that are not related to SARS-CoV-2. Additionally, disease-induced deaths are included in the model with the aid of the 2 mortality rates λ_1 and λ_2 that are explained in their context and defined in the following subsections. As explained in Section 2.5 it may cause too large numbers of corona-induced deaths in the end if data sources predefine deceased corona-infected individuals as SARS-CoV-2 deaths.

(B2) Modelling the spread of and measures against 2019-nCoV in any European state in general and Germany and Sweden in particular, a system inflow rate describing new entering into the class S can be included. An epidemic or pandemic, where individuals newly enter the system, is also called an *open infectious disease* [147, p. 18]. For various diseases the system inflow rate is strongly influenced by the number of births into the system per unit of time. Regarding the novel corona virus, where the occurrence of infection over several months and not years is observed, birth rates can be regarded as negligible. The system inflow rate can instead be represented by a so-called *recruitment rate* that is the rate at which for instance tourists and immigrants enter the system. For the first, this means that a fixed number of individuals newly flows into the system per unit of time. The factors influencing recruitment of susceptibles have to be selected in this case.

With regard to extremely reduced travel activities during the corona virus pandemic and an assumed general balance between tourists entering and leaving the country per unit of time the recruitment rate can be omitted. If no tourism and no births are taken into account and the individuals in the regarded population can die and thus leave the system, the total number of individuals is not assumed to be a constant N at all time instances, but described by a time-dependent function $N(t)$. On the one hand, the usage of a standard incidence, in which the number of susceptibles at a time instant t $S(t)$ is normalized by the total population size

$N(t)$, requires the computation of the time derivative of the time-dependent relation $\frac{S(t)}{N(t)}$ by application of the quotient rule of differential calculus. On the other hand, constancy can be achieved by adding the value $\mu \cdot N$ as a recruitment rate to the ODE related to the susceptible compartment.

(B3) In the established model, a certain equivalence to the presupposition of a permanent cure as in the Basic SIR model in Chapter 3 is present. In Section 4.2.9, it is explained that and why individuals are not assumed to be able to move from the recovered to the susceptible compartment in the SARS-CoV-2-fitted model.

Possibilities of incorporating a waning protective effect are detailed. Section 4.2.11 states in which ways a waning effect can be incorporated into a model with a considered vaccination program.

(B4) Individuals infected with the novel corona virus are not automatically infectious. Therefore, a latent and an incubation period are considered in the SARS-CoV-2-fitted model. They are pictured in Section 4.2. However, in the implemented model variants, which will be introduced in Subsection 5.2.2, the group of infected people in the population is not subdivided into several compartments. This is reasoned by unavailable or non-existent data reports referring to the number of latently or pre-symptomatically infected cases.

4.2   Compartments and Transitions in the SARS-CoV-2-fitted SIR Model

The most relevant feature of the SARS-CoV-2-fitted model is the separation of the infected and infectious compartment I of the Basic SIR model into eight classes, each of which symbolizes a specific stage of disease progression. This division emerges from the model extensions presented in Subsection 3.2. For this reason, two exposed, two asymptomatic compartments and a hospitalized as well as an intensive care unit compartment are integrated into the model.

Furthermore, the control strategies of quarantine, isolation as well as vaccination are considered in the model. A deceased compartment comprises all of the individuals that die for reasons connected with or caused by a SARS-CoV-2 infection. The definition of a COVID-19 death as well as the inclusion of deaths potentially related to COVID-19 may differ between countries, regions and institutions as mentioned in detail in Section 2.5.

Besides, it is significant to determine the effects of testing involving positive results for the transitions in the model. Clearly, testing is possible in all compartments except for the one comprising disease-induced deaths, and comprises approved, authorized and applied SARS-CoV-2 tests that lead to a case-confirmation. In the case of an assumed complete immunity testing is not performed if a person has recovered and so transited to the recovered compartment. Testing is not performed if a person is already a known confirmed case.

Positively tested individuals are defined as *confirmed cases* in the SARS-CoV-2-fitted model.

Section 4.2.1 provides a definition of compartment sojourn times, which is used to define transition rates in Subsections 4.2.3 to 4.3.11. Each of these nine subsections deals with a population group characterized by specific features with respect to infectedness, infectiousness, transmissibility and disease progression, and is classified as a compartment of the emerging model or subdivided into sub-compartments.

In Subsection 4.2.2, functions describing the probability distributions of the incubation period and the serial interval with regard to COVID-19 are explained as alternatives to the commonly used exponential distribution. Subsection 4.2.12 gives an overview of the characteristics of the introduced compartments.

4.2.1   Sojourn Times

The object of primary interest with respect to the sojourn time of individuals in a compartment is the *survival function* $\mathcal{G}(t)$, which is the probability that the time of leaving the current compartment is later than some specified time t. It is defined by [202, p. 541]

$$\mathcal{G}(t) = \mathbb{P}(t_D > t) = 1 - \mathcal{R}(t) \, , \mathcal{G}(0) = 1 \, ,$$

where $t \geq 0$ is some time, t_D is a random variable denoting the time to switching to the next compartment, and $\mathcal{R}(t) = \mathbb{P}(t_D \leq t)$ is the so-called *reliability function* [203]. If $p(t)$ is the PDF concerning the cumulative distribution function $\mathcal{R}(t) = \mathbb{P}(t_D \leq t)$, the average time of sojourn T_S in the current infected state is given by [202, p. 541]

$$T_S := \int_0^\infty t \cdot p(t) dt \, . \tag{4.1}$$

If it is assumed that $\lim_{t \to \infty} t \cdot \mathcal{G}(t) = 0$, Equation (4.1) becomes [202, p. 541]

$$T_S := \int_0^\infty \mathcal{G}(t) dt \tag{4.2}$$

If we have for instance an exponential PDF $p(t) = \frac{1}{\mathcal{D}} \cdot e^{-\frac{t}{\mathcal{D}}}$ with $t \geq 0, \mathcal{D} > 0$, then it holds that $\mathcal{G}(t) = e^{-\frac{t}{\mathcal{D}}}$ such that

$$\int_0^\infty \mathcal{G}(t) = \int_0^\infty e^{-\frac{t}{\mathcal{D}}} dt = \mathcal{D} = \int_0^\infty \frac{t}{\mathcal{D}} \cdot e^{-\frac{t}{\mathcal{D}}} dt = \int_0^\infty t \cdot p(t) dt . \qquad (4.3)$$

So the average time of sojourn is $\mathcal{D}$ units of time in this case. It should be noted that Equation (4.3) leads to the same assumption with respect to the mean sojourn time in a compartment as outlined in Equation (3.5) in Subsection 3.1.3 with reference to the Basic SIR model. The average duration of stay in a compartment is used to describe the rate at which individuals transit to the respective next compartment in the following subsections of Section 4.2.

4.2.2   Distributions of the Incubation Period and the Serial Interval

Time-deterministic epidemic models are often formulated as systems of ODEs with an exponentially distributed time of stay in each infected compartment. A classical extension to account for more realistic distributions of incubation and infectious periods concerning the stay of individuals in compartments is to split infected compartments into distinct stages [78]. Generally, different stages of disease progression, like a latently infected, carrier, pre-symptomatic, symptomatic or seriously affected stage can be used in order to model the incubation period and the serial interval of SARS-CoV-2 as random numbers following certain probability distributions. In the following approach based on a deterministic SIR model, the infected compartment I is divided into multiple infected classes.

Let there be j infected states $I_1, I_2, ..., I_{k-1}, I_k, I_{k+1}, ..., I_j$, in the SARS-CoV-2-fitted model, of which $I_{k+1}, ..., I_j$ are contagious to susceptibles as well as symptom-exhibiting. A compartment model resulting from such subdivisions is a classical extension of the standard disease transmission model or Basic SIR model to account for non-exponential distributions of incubation and infectious periods [78].

The average lengths of stay in the compartments I_1 to I_k account for the incubation period $\mathcal{T}^I$. The average length of stay in each of the compartments I_{k+1} to I_j does not contribute to the incubation period. In the SARS-CoV-2-fitted model presented in the following sections, the $j = 5$ states of latent, asymptomatic infectious, symptomatic, hospitalized and intensive care infectedness are introduced, of which exclusively the latently infected and asymptomatic infectious state are definitely not symptomatic, such that $k = 2$. Of course the latently infected and asymptomatic contagious compartment can both be subdivided into more classes with distinct characteristics.

Let the infected states I_i for all $i \in \{1, ..., k\}$ be random numbers that are distributed according to the exponential distribution with a *rate parameter* λ [204]. In the context of transitions between infected compartments, $I_i \sim \text{Exp}(\lambda)$ means that λ is the number of expected events (=number of compartments that an individual passes within a unit of time). An individual passes a single compartment within $\frac{T^I}{k}$ days, and $\lambda = \frac{k}{T^I}$ compartments within one day.

It holds that [205]

$$I_i \sim \text{Exp}(\lambda) \quad \forall i \in \{1, ..., k\} \Rightarrow \sum_{i=1}^{k} I_i \sim \text{Erl}(\lambda, k) .$$

This means that the sum over k random numbers, that are all exponentially distributed according to the parameter λ, is distributed according to the *Erlang distribution* with the parameters λ and k. [204] The parameter k is called the *shape parameter*. The Erlang distribution is a continuous probability distribution.

Let $X \sim \text{Erl}(\lambda, k)$. The PDF of the Erlang distribution is given by [206]

$$f_{\text{erlang}}(t) = \begin{cases} \frac{\lambda^k \cdot t^{k-1}}{(k-1)!} \cdot e^{-\lambda \cdot t} & \text{if } t \geq 0 \\ 0 & \text{if } t < 0 , \end{cases}$$

where $k \geq 1$ and $\lambda > 0$. For an Erlang-distributed random variable X, the probability of lying in the interval $[0, t]$ is given by the Erlang cumulative distribution function [206]

$$F_{\text{erlang}}(t) = \begin{cases} 1 - e^{-\lambda t} \sum_{m=0}^{k-1} \frac{(\lambda \cdot t)^m}{m!} & \text{if } t \geq 0 \\ 0 & \text{if } t < 0 . \end{cases}$$

This is the cumulative distribution function of the Erlang distribution. The expected value and variance of the Erlang distribution are [207]

$$\mathbb{E}[X] = \frac{k}{\lambda} = T^I, \quad V[X] = \frac{k}{\lambda^2} = \frac{T^{I^2}}{k} .$$

The function $f_{\text{erlang}}(t)$ yields the probability for the occurrence of the kth event if $\lambda = \frac{k}{T^I}$ events are expected per unit of time. Because $\text{Erl}(\lambda, 1) \equiv \text{Exp}(\lambda)$ i.e. the Erlang distribution with shape parameter $k = 1$ is equivalent to the exponential distribution, the Erlang distribution is a generalization of the exponential distribution. It is also a special case of the Gamma distribution since it equals the Gamma distribution with an inverse scale parameter λ and a shape parameter k. Moreover, it is the conjugated distribution of the Poisson distribution [206]. The Erlang PDF is depicted in Figure 4.1 for an average incubation period of 6 days and 1, 4 or 6 intermediate infected states represented by compartments lying between the susceptible and the symptomatic infectious class. One unit of time is assumed to be one day here.

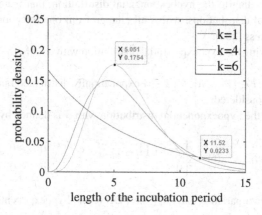

Figure 4.1 Erlang probability density function with $\mathcal{T}^I = 6$ and $k = 1$ or $k = 4$ or $k = 6$.

Figure 4.1 conveys that the PDF of the Erlang distribution with a shape parameter $k = 1$ equals the PDF of the exponential distribution. In the case of a shape parameter allocation of $k = 4$, which means 4 intermediate infected states, 4.444 days of incubation, and with $k = 6$ intermediate infected states 5.051 days of incubation time are necessary to achieve the highest possible value of probability density. For a fixed value of $\mathcal{T}^I$, more intermediate infected states result in a shorter time of stay in every single infected class and a larger number of infected compartments passed within a single day. This effects a higher maximal probability density.

All infectious compartments can induce a transmission and so a secondary infection caused by a contact of an infectious individual with a susceptible person. The average beginning of a secondary infection after the initiating infection marks the length of the serial interval. Therefore, the serial interval can be modelled by the average interval from contagion until loss of infectiousness.

It is now assumed that a rate ω exists at which individuals transit from the compartment I_{k+1} to the recovered compartment R, taking into account a direct transition or a transition over all or some of the compartments $I_{k+2}, ..., I_j$ [204]. Let $I_{k+1} \sim \text{Exp}(\omega)$. It still holds that $I_i \sim \text{Exp}(\lambda) \ \forall i \in \{1, ..., k\}$. Consequently, the probability distribution of the sum of the random variables $I_1, ..., I_{k+1}$ is given by the *hypoexponential distribution* [208]:

$$\sum_{i=1}^{k+1} I_i \sim \text{Hypoexp}(\alpha)$$

with $\alpha = \{\frac{k}{TI}, ..., \frac{k}{TI}, \omega\}$, where $\frac{k}{TI}$ appears k times. Hence, the serial interval is distributed according to the hypoexponential distribution, that is a series of multiple exponential distributions, which all emerge from one infectious state of the incubation process [204].

In the following the hxpoexponential distribution with $\tilde{\alpha} = \{\tilde{\alpha}_1, \tilde{\alpha}_2, ..., \tilde{\alpha}_n\}$, $n \in \mathbb{N}$,

$\tilde{\alpha}_i \neq \tilde{\alpha}_l$ $\forall i, l \in \{1, ..., n\}$, for $k+1$ exponentially distributed random numbers $I_1, ..., I_{k+1}$ is considered.

The PDF of the hypoexponential distribution with $\tilde{\alpha}$ is defined by [209]:

$$f_{\text{hypoexp}}(t) = \begin{cases} \sum_{i=1}^{k+1} \prod_{l \neq i} \frac{\tilde{\alpha}_l \cdot \tilde{\alpha}_i}{\tilde{\alpha}_l - \tilde{\alpha}_i} \cdot e^{-\tilde{\alpha}_i \cdot t} & \text{if } t > 0 \\ 0 & \text{if } t \leq 0 \end{cases}$$

In the case of two parameters $\tilde{\alpha}_1, \tilde{\alpha}_2$ the PDF of the hypoexponential function is defined by

$$f_{\text{hypoexp}}^{(n=2)}(t) = \frac{\tilde{\alpha}_2 \cdot \tilde{\alpha}_1}{\tilde{\alpha}_2 - \tilde{\alpha}_1} \cdot e^{-\tilde{\alpha}_1 \cdot t} + \frac{\tilde{\alpha}_1 \cdot \tilde{\alpha}_2}{\tilde{\alpha}_1 - \tilde{\alpha}_2} \cdot e^{-\tilde{\alpha}_2 \cdot t} = \frac{\tilde{\alpha}_1 \cdot \tilde{\alpha}_2}{\tilde{\alpha}_1 - \tilde{\alpha}_2} \cdot \left(e^{-\tilde{\alpha}_2 \cdot t} - e^{-\tilde{\alpha}_1 \cdot t} \right).$$

The definition of the respective cumulative distribution function is [210]

$$F_{\text{hypexp}}(t) = \begin{cases} 1 - \sum_{i=1}^{k+1} \prod_{l \neq i} \frac{\tilde{\alpha}_l}{\tilde{\alpha}_l - \tilde{\alpha}_i} \cdot e^{-\tilde{\alpha}_i \cdot t} & \text{if } t > 0 \\ 0 & \text{if } t \leq 0 \end{cases}$$

The origin of the name of the hypoexponential distribution is the fact that the coefficient of variation $\frac{\sqrt{\alpha_1^2 + \alpha_2^2}}{\alpha_1^2 + \alpha_2^2}$ (ratio of the standard deviation to the mean) is smaller than 1, as 1 is the coefficient of variation of the exponential distribution [211]. The hypoexponential distribution is also called the generalized Erlang distribution.

If $X \sim \text{Hypoexp}(\tilde{\alpha})$ and in the 2-parameter case, the expected value and variance are given by

$$\mathbb{E}[X] = \frac{1}{\tilde{\alpha}_1} + \frac{1}{\tilde{\alpha}_2} \, , \, Var[X] = \frac{1}{\tilde{\alpha}_1^2} + \frac{1}{\tilde{\alpha}_2^2}.$$

In Figure 4.2 the PDF of the hypoexponential distribution for $\tilde{\alpha} = \{\lambda, \omega\}$, an average incubation period of 6 days and 1 or 3 intermediate infected states as well as a medium time to the loss of infectiousness of 5 days is depicted.

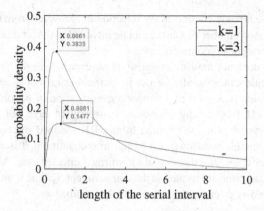

Figure 4.2 Hypoexponential probability density function with $\omega = 5$, $T^I = 6$ and $k = 1$ or $k = 3$.

Figure 4.2 shows that the length of the average time period between an infection of an individual and an emerging secondary infection of less than 1 day is necessary to achieve the highest possible value of probability density supposing parameter values of $\omega = 5$, and $k = 3$ or $k = 1$. The probability density reached with 1 intermediate infected state is around 0.23 smaller than with 3 intermediate states. For a fixed value of T^I, more intermediate infected states lead to an increase in the maximally attainable probability density.

It is worthwhile noting that both the Erlang and the Hypoexponential distribution yield models for inter arrival times or service times in queuing systems. The Erlang distribution is a phase-type distribution where the k phases are in series, that have the same parameter for the exponential distributions. The hypoexponential distribution is a phase-type distribution where n phases are in series and the phases have distinct exponential parameters [212]. It has to be mentioned that the incubation period or serial interval could be modelled with the aid of a different distribution than the Erlang or Hypoexponential distribution, respectively, but the two introduced probability distributions yield reasonable results in modelling the 2 intervals.

4.2.3   The Susceptible Compartments

The compartment S contains the susceptibles of the population. Concerning 2019-nCoV, these are all inhabitants of the regarded country or region except for those people who have already contracted the infection or recovered from the disease. A constantly assumed number of tourists can be included in the total population, too, without incorporating a recruitment rate.

The class S does not have to comprise all susceptibles in the system because susceptible people can generally be put in quarantine including self-quarantine. Inflicted quarantine is defined as a temporary segregation of people suspected of infectedness [213]. In Germany, quarantine is usually 14 days, but 10 days if the respective susceptible has just returned from a RKI-defined risk area [214]. For example, a susceptible individual can be put in quarantine if it is registered as a contact person or household member of a confirmed infected case. All susceptibles in preventive quarantine are located in the compartment S_q. The transition from the class S_q back to S is given by the rate $\hat{\phi}$, which is defined as

$$\hat{\phi} = \frac{1 - \mu \cdot Q}{Q},$$

if the average length of quarantine is Q. It is significant to notice that all individuals in the class S_q are assumed not to be exposed to the infection risk due to their quarantine, and are assumed to receive a negative test result if they are tested. The quarantine rate q has some influence on the rate $\theta_{S_q}(t)$, that describes the transition from S to S_q and will be explained in Section 4.3. It can be altered in the later implementation for the purpose of comparing scenarios of different quarantine measure up to a lockdown with each other. This is performed in Section 6.3 in Chapter 6.

Apart from the compartment S_q individuals can progress to the exposed classes E or E_I from S. The exposed compartments are explained in the next subsection. The compartments S_q, E and E_I are the only 3 ones accessible for susceptibles.

It is important to notice that a certain number of individuals can remain in the class of susceptibles forever or at least for a long time. A transition between S and S_q is always possible.

The German RKI recommends a quarantine of 14 days for susceptibles who have just had a contact of category I with a confirmed infected case. The RKI defines a close contact that is longer than 15 minutes, with less than one and a half metres distance between the contact persons and no correct continuous wearing of face masks, or took place in a room with a probably high concentration of infectious aerosols and took longer than 30 minutes, as a *contact of category I*, meaning with a higher infection risk for the susceptible [215]. The quarantine can be shortened to 10 days if a negative PCR or antigen test is obtained from the respective person [216].

The ECDC defines a *close contact person* as an individual that has had face-to-face contact with a COVID-19 case within 2 metres for more than a total of 15 minutes over a period of one day, even if not consecutive, had physical contact with a COVID-19 case, had direct contact with infectious secretions of a COVID-19 case, was in a closed environment or travelling with a COVID-19 case for more than 15 minutes, is a person providing direct care to a COVID-19 case, a laboratory worker handling specimens from a COVID-19 case without recommended personal protective equipment or with a possible breach of personal protective equipment or hand hygiene [217]. A contact with only a low risk of contagion is defined as a person who has had face-to-face contact with a COVID-19 case within 2 metres for less than 15 minutes, was in a closed environment or travelling with a COVID-19 case for less than 15 minutes [217]. Household contacts and contacts around the time of symptom development of the infected person are generally regarded as putting the susceptible contact person at an increased risk [217]. The ECDC describes the main principles or contact tracing as the prompt identification of contacts of COVID-19 cases, the provision of self-quarantine, proper hand hygiene and respiratory etiquette, testing all high-risk exposure contact persons as well as low-risk exposure contact persons in settings in which transmission is likely as well as testing of all contacts that become symptomatic [217].

Moreover, an US-American study quantified levels of SARS-CoV-2-reactive antibodies and human corona virus-reactive antibodies in serum samples collected from 431 humans before the COVID-19 pandemic and pre-pandemic antibody levels in serum from a separate cohort of 251 individuals who became PCR-confirmed infected with SARS-CoV-2. It indicated that pre-pandemic non-neutralizing antibodies elicited by 196 human corona viruses did not provide SARS-CoV-2 protection [218]. Thus it also implied that a previous infection with another corona virus did not protect from an infection with SARS-CoV-2. So all population members can be assumed to be initially susceptible in the model.

4.2.4   The Exposed Compartments

The first step of a model extension apart from the introduction of a quarantine rate is the extension of the Basic SIR model to a $SEIR$ model. As in the $SEIR$ model the exposed class E comprises latently infected people, who are not yet infectious or show symptoms.

With respect to SARS-CoV-2, a distinction between confirmed and unconfirmed infected cases has to be made. Any type of data source can only provide a reliable SARS-CoV-2 incidence on the basis of the numbers of infected cases that have been reported to any kind of authority until then.

In the SARS-CoV-2-fitted model, a difference is made between confirmed and unconfirmed cases. An exposed compartment E solely contains unconfirmed infected cases.

Another exposed class E_I comprises those latently infected individuals who are confirmed cases. To be able to obtain reliable compartment size data, it is presupposed that **a test realization triggers isolation imposture on latently infected individuals.** Therefore, an isolation rate $\mathcal{I}$ is introduced, which has some influence on the transmission rate, that is explained in Section 4.3. Solely a received positive test result is assumed to effect a transition to the compartment E_I. In Germany, confirmed cases are isolated at home for 10 days [213]. The RKI distinguishes between cases that are reported owing to laboratory diagnostics, which usually means a SARS-CoV-2 test, and cases that are reported to the authorities for clinically-epidemiologically verified reasons [219]. However, an entailed test is necessary in the second case for a case-confirmation.

It is assumed that a latently infected individual that is not positively tested (situated in the compartment E) is not isolated. This is justifiable since the detection of infected cases who are asymptomatic and not infectious is not effected by symptom development, which is a common trigger for isolation.

Incorporating the compartments E and E_I into the model, all latently infected cases that are not confirmed and not isolated as well as all latently infected cases that are confirmed and isolated are included.

If an individual receives a false-positive test, it is not infected but isolated, but still progresses to the compartment E_I. False-negative tests are assumed to be rare. If a person receives a false-negative test, it is infected but not isolated, but still transits to the compartment E. Its infection can be confirmed later such that it will transit to a compartment that is distinguished by confirmed infectedness.

The rate describing the transmission affecting the transition from S to E or E_I, respectively, is a time-dependent continuous functions $\theta_E(t)$ or $\theta_{E_I}(t)$. The transmission risk β, the contact rate function $\gamma(t)$ and the quarantine rate function $q(t)$ as well as the isolation rate $\mathcal{I}$ influence them. The compositions of $\theta_E(t)$ and $\theta_{E_I}(t)$ are explained in Section 4.3, in which the transmission in the SARS-CoV-2-fitted model is derived.

4.2.5   The Asymptomatic Infectious Compartments

Six additional classes are added to the extended model for the purpose of reflecting different infected states that are passed by individuals during an infection.

There are two compartments A and A_I which contain infectious but pre-symptomatic individuals.

Exposed unisolated individuals transit from the compartment E to the infectious, pre-symptomatic class A at a rate ψ_U. The pre-symptomatic class is named the asymptomatic class in throughout this thesis. Latently infected individuals who are in isolation (E_I) will spend their whole disease progress in isolation and will not become unconfirmed cases. For this reason, there is no possibility for them to reach the compartment A of unconfirmed asymptomatic individuals anytime. They progress to the isolated stage A_I at a rate ψ_{II}. Individuals who are latently infected but not isolated (E) are able to progress to the class A_I instead of A at a rate ψ_I if they are detected and reported to the authorities.

The number of individuals in the isolated class A_I is relatively small compared to the number of people in A because most often the testing of a case is triggered by symptom development.

The mentioned transition rates ψ_U, ψ_I and ψ_{II} depend on the length of the latent period $\mathcal{L}$. Their composition and the option to model them by time-dependent functions by creating a time delay model based on *Delay Differential Equations* (DDEs) are explained in Subsection 4.3.4.

4.2.6   The Symptomatic Compartments

The compartments I_U and I_I are both composed of infected symptomatic infectious individuals only. The class I_I contains those cases that have been confirmed by responsible authorities like health departments and are thus confirmed as infected cases. Individuals in the class I_I are isolated at home and may be sent to hospital after a certain time period if they exhibit major symptoms. The compartment I_U consists of the respective unconfirmed cases.

Placed in the asymptomatic compartment A_I, individuals transit to the compartment I_I at a rate χ_{II}, which depends on the length of the incubation period $\mathcal{T}^I$ and the length of the latency period $\mathcal{L}$. The length of the time period between the development of contagiousness and the symptom start τ is defined as

$$\tau = \mathcal{T}^I - \mathcal{L} \, .$$

It is certain that all individuals in the class A_I will reach I_I because they definitely pass through the incubation time. **All infected individuals are assumed to show at least mild symptoms for the duration of the disease in this model, such that pauci-symptomatic people are counted as symptomatic, too.**

Asymptomatic cases that are not in isolation (A) have not yet been confirmed by the authorities. For this reason, they may remain unconfirmed infected cases during their whole infected and infectious process and transit to the compartment I_U at a rate χ_U.

It is not assumed that individuals in the compartment A_I can arrive at the class I_U at any time as their infected status is already known. Since the time instant when their test result was positive and they transited from S to E_I or from E to A_I, they have been confirmed cases.

People are infectious in the class A and their transition to the symptomatic stage of infection is initiated when they develop symptoms. As mentioned before, symptom development is a common occasion for reporting an infection. It is assumed to be probable that individuals visit a doctor and/or a testing station and are tested, or at least report their infected status, as soon as their first symptoms appear. Subsequently, these individuals turn from unconfirmed into confirmed infected cases.

It follows that there is a possibility for individuals in A to reach the class I_I instead of I_U at a rate χ_I and belong to the confirmed symptomatic compartment as soon as the incubation time has passed for them. The rate χ_I depends on the testing and entailed reporting of the respective infection.

Obviously, asymptomatic individuals that are not isolated, who do not transit to the class I_I, reach the compartment I_U and vice versa. Thus the 2 rates χ_I and χ_U can be described as *coupled*. The ratio of asymptomatic individuals who turn into confirmed cases via testing at symptom-development among all asymptomatic people who have been unconfirmed up to the respective time instant is given by the parameter κ_I. It is called a *detection parameter* here. The corresponding ratio of asymptomatic individuals, who remain unconfirmed, is then given by the parameter $\kappa_U = 1 - \kappa_I$.

The same holds for the 2 transition rates ψ_I and ψ_U. Here the detection parameter ζ_I is the ratio of exposed individuals who are confirmed by a testing procedure among all exposed cases who have been unconfirmed up to the respective time instant. The corresponding ratio of exposed individuals who remain unconfirmed is described by the parameter $\zeta_U = 1 - \zeta_I$.

The parameter J denotes the ratio of the unconfirmed symptomatic cases that are assumed to be confirmed by test realization during their period of symptomatology per time instant t. Let T be the average length of the time that individuals in I_U and I_I take from the symptom development to recovery.

It is significant to add a transition from the unconfirmed symptomatic infectious compartment I_U to the deceased compartment D. The reason is that those individuals who have never been tested are no confirmed SARS-CoV-2-cases, such that a death of a person in the compartment I_U is usually traced back to a comorbidity, or respiratory disease, that has actually been caused by a SARS-CoV-2 infection. It has been explained in Chapter 2 why a lack of test realizations can lead to an overestimation of the overall CFR.

Let M_U be the CFR of unconfirmed infectious cases.

Altogether, the individuals in I_U are assumed to transit to the recovered class R at a rate

$$\omega_1 = \frac{1 - \mu \cdot T}{T} \cdot \left(1 - J - M_U\right)$$

and to the compartment I_I at a rate

$$\zeta = \frac{1 - \mu \cdot T}{T} \cdot J .$$

The rate at which individuals transit from I_U to D is described by the parameter

$$\lambda_3 = \frac{1 - \mu \cdot T}{T} \cdot M_U .$$

Furthermore, let K be the case-hospitalization rate of symptomatic infectious individuals. The compartment H of hospitalized cases can only be reached from the class I_I. Unconfirmed cases in I_U are expected to never be hospitalized as SARS-CoV-2-patients because hospitalized cases are always confirmed but individuals in I_U have not been (positively) tested and remain unconfirmed.

Since the individuals in I_I are assumed to either recover after T units of time or be hospitalized owing to major symptoms they progress to the compartment R at a rate

$$w_2 = \frac{1 - \mu \cdot T}{T} \cdot \left(1 - K\right)$$

and to the compartment H at a rate

$$\eta = \frac{1 - \mu \cdot T}{T} \cdot K .$$

It is highlighted at this point that it is assumed in this model that all transitions from the susceptible or an unconfirmed infected compartment to a confirmed infected class are triggered by a positive SARS-CoV-2 test. This concerns the transitions from an exposed unconfirmed to a confirmed infectious, or unconfirmed asymptomatic to a confirmed symptomatic, or susceptible to an exposed compartment.

A test that entails a transition from the compartment A to I_I is more probable to be effected by case-reporting than a transition from S to E_I because symptom development is the most obvious reason for testing.

4.2.7   The Hospitalized Compartment

The seventh infected model compartment is the class H of all patients that are hospitalized due to the consequences of their SARS-CoV-2 infection or secondary infections.

Let M_H be the CFR of the hospitalized compartment. Let T_H be the average time period from hospitalization to recovery. A fraction ι of hospitalized patients is assumed to be transferred to intensive care per unit of time. Individuals transit from the hospitalized compartment to the ICU compartment C at a rate

$$\xi = \frac{1 - \mu \cdot T_H}{T_H} \cdot \iota \, ,$$

decease for reasons related to SARS-CoV-2 at a rate

$$\lambda_1 = \frac{1 - \mu \cdot T_H}{T_H} \cdot M_H$$

and recover at a rate

$$\omega_3 = \frac{1 - \mu \cdot T_H}{T_H} \cdot \left(1 - \iota - M_H \right) .$$

4.2.8   The Intensive Care Unit Compartment

Hospitalized SARS-CoV-2-cases can be transferred to an ICU. It makes sense to include a correspondent intensive care compartment C in the model because the number of future patients in intensive care can be predicted like that. The letter C stands for critical cases. Additionally, the number of daily new patients in intensive care is more certain than other data. This is further detailed in Section 5.4.

In this thesis, the CFR of intensive care patients is assumed to be a multiple of the CFR of the hospitalized compartment. To obtain the CFR of C, the CFR of H is multiplied by a factor $x_\iota > 0$ that enhances the case-fatality rate of ICU patients on the basis of the CFR of hospitalized patients who are not in an ICU:

$$M_\iota = x_\iota \cdot M_H > M_H.$$

Moreover, let the average time from admission to intensive care to recovery be $T_\iota > T_H > T$.

Individuals who are in intensive care die for reasons related to SARS-CoV-2 at a rate

$$\lambda_2 = \frac{1 - \mu \cdot T_\iota}{T_\iota} \cdot M_\iota$$

and recover at a rate

$$\omega_4 = \frac{1 - \mu \cdot T_\iota}{T_\iota} \cdot \left(1 - M_\iota\right).$$

4.2.9   The Recovered Compartment

In this compartment model, recovery of a person is identified with the achievement of an amount of viruses in the individual that makes it impossible or exceedingly improbable to transmit the infection to susceptible people. Recovery is not equalled with the complete loss of contagiousness here, since the time period until infectiousness fully vanishes is difficult to determine, and the time period until a person reaches a level making it not infectious to others is more relevant. Recovery is also not identified with the disappearance of symptoms in this thesis. This is significant since individuals who have transited to the compartment R in the SARS-CoV-2-fitted model might have lost remarkable symptoms for some time already but not their complete infectiousness. They should not be able to infect susceptibles anymore if they are classified as recovered.

The average time to recovery of the compartments I_I and I_U is T, of the class H is T_H and of the class C is T_ι in this model. Thus the average length of the infectious period has to be determined in order to equip the model with an adequate value of T. The average time from hospitalization or ICU admission to recovery is used to determine the values of T_H and T_ι. As mentioned, it is not yet clear how long these periods exactly are [10]. It is however verified that contagiosity reaches its maximum at the time when people develop symptoms, which is the time when the incubation period ends. It is also assured that infectiousness of an individual decreases in the course of time, and severely diseased individuals are infectious for a longer time than people with mild symptoms [10]. The values for T, T_H and T_l are obtained from the data provided by the RKI [10]. The RKI states for example that infectiousness intensely declines after 10 days if a person exhibits only mild symptoms [10].

With respect to recovery from COVID-19, it has to be considered that in reality, the duration of resistance against a disease after overcoming it naturally or after vaccination can vary over time [220]. According to current information, it might be possible that symptoms re-appear at any point of time and even susceptibility might be reached again by (some) individuals due to a waning protective effect of recovery. The relationship between the presence of antibodies to severe SARS-CoV-2 and the risk of subsequent re-infection currently remains unclear [221]. The period until a critical immunity threshold is fallen below could be mathematically modelled by a probability distribution in future research approaches.

In general, waning immunity might cause recurrent outbreaks of infectious diseases [222]. A missing complete immunity or a waning immunity effect can be incorporated into the model in different ways. Only a few approaches to compart-ment models with waning immunity can be found in the literature [220]. Examples are DDE models, which can for instance account for a delay representing the average duration of the diseases-induced immunity [222]. They are introduced in Subsec-tion 4.3.4. Models of partial differential equations (PDEs) are often used to include pathogen transmission among distinct age groups [222]. The loss of immunity after recovery or vaccination can be involved in a model by including a second suscepti-ble compartment, that comprises individuals who are excluded from the possibility of getting vaccinated since nobody is aware that their immunity level has dropped below a critical threshold, as proposed by Barbarossa et al. [222].

Another option is that individuals enter the susceptible compartment as soon as they are not protected from contracting the infection due to their previous infection and recovery or vaccination anymore. This is realized in the SIR model introduced by Dafilis et al. [223], that has an intermediate waning state and makes damped or undamped oscillations in the system observable.

A transition from the class R back to S would also be justifiable if fully recov-ered individuals were able to get infected again i.e. contract a mutated form of the virus. This is not taken into account in the established model since the risk of re-infection has not been fully proved with regard to all known mutations, that generally exhibit a higher transmissibility. This is certainly a model simplification and does not apply to all individuals, substantially if an increasing occurrence of more aggressive mutations is assumed. Nonetheless, if a re-infection with mutations is possible, it can be assumed that the amount of re-infections leading to major symptoms, severe disease progressions and death has amounted to only a small fraction of a population of several millions of people until today. The precise inclusion of different mutations in the model is relevant for future work.

A carrier compartment between the classes R and E would be reasonable if a noticeable number of people in the system did not fully recover after their initial infection so that they were able to infect susceptibles again. In a longitudinal cohort study the incidence of SARS-CoV-2 infection confirmed by PCR in seropositive and seronegative health care workers attending testing of 12,541 people, who were asymptomatic and symptomatic staff at Oxford University Hospitals, was investigated using an anti-trimeric spike immunglobulin G (IgG) enzyme-linked immunosorbent assay (ELISA). Baseline antibody status was determined by anti-spike and anti-nucleocapsid IgG assays, and staff members were followed for up to 31 weeks. No symptomatic infections and only 2 PCR-positive results in asymptomatic workers were seen in those with anti-spike antibodies, which suggests that previous infection resulting in antibodies to SARS-CoV-2 is associated with a protection from re-infection for most people for at least 6 months [221]. As consequence of this and the fact that compartment sizes are predicted over up to the future 80 weeks (6.67 months) in Chapter 6, it is assumed in this model that the amount of viruses in cases with re-appearing symptoms does not attain a level that makes them contagious to others again. Hence, there is no carrier compartment or possible transition from the compartment R to E.

Carriers in the compartment I_I or I_U who do not show relevant symptoms anymore but are still infectious are possible. No additional carrier compartment is added since it is unclear which proportion of again asymptomatic individuals among the infectious people is. Also, symptoms play a role subordinate to infectiousness in the model. Reduced contagiousness in individuals in the compartments H and C can occur prior to the reduction of symptoms, or the loss of symptoms prior to a decreased infectiousness. This is already implied by the fact that infectiousness and symptoms are no characteristic features of the hospitalized or ICU compartment in this thesis.

As indicated above, the rates at which R is reached by unconfirmed symptomatic cases is ω_1, tested symptomatic cases is ω_2, hospitalized patients who are not in intensive care is ω_3, and intensive care patients is ω_4. Once infected, so once arrived in the compartment E or E_I, individuals certainly reach the recovered compartment R after some time, unless they die.

4.2.10 The Compartment of Deceased Individuals

The infected people in H or C underlie certain time-dependent case-fatality rates M_H or M_l, respectively, which make them reach the deceased compartment D at a rate λ_1 or λ_2, respectively. Hence the deceased compartment $D(t)$ contains all individuals who have been classified as deceased SARS-CoV-2 deaths up to the time instant t by the data source that is selected for the implementation. If IFRs from reliable sources are available, they can be used instead of the CFRs, which are provided by most reliable sources, in order to reflect the lethality rate of the infectious disease in the most realistic way.

4.2.11 The Compartment of Vaccinated Individuals

Subsection 3.2.4 implied the possibility of the integration of vaccination into the model. It is assumed that only susceptible individuals are vaccinated in the SARS-CoV-2-fitted model. Vaccinations of confirmed infected cases are not realistic, and vaccinations of undetected latently infected or asymptomatic infectious are omitted here because they represent only a small portion of vaccinations. The STIKO recommends that (previously) infected individuals should be vaccinated soonest 6 months after recovery [224]. As it can be assumed that recovered individuals subsequently accounted for only a negligible fraction of vaccinated individuals as of March 2021, which is the point in time up to which compartment size data is available in this thesis, vaccinations of recovered people are omitted here.

It must be decided whether a considered SARS-CoV-2 vaccine is a so-called ”*leaky*” or ”*all-or-nothing vaccine*”. If an all-or-nothing vaccine is assumed, a vaccination provides complete protection from the infection for a fraction $\mathcal{V}_1$ of the susceptible class per unit of time t, whereas the fraction $1 - \mathcal{V}_1$ does not gain any protection. Booster vaccinations after several years are still possible in this case in order to maintain vaccine protection. Susceptible individuals who receive a vaccination transit from the compartment S to a vaccinated compartment V at the rate $\mathcal{V}_1$. This rate can be defined as a time-dependent function $\mathcal{V}_1(t)$ if a fluctuating vaccine strategy applied to the population or an increasing number of available doses is assumed. Therefore, the compartment $V(t)$ comprises all of the individuals in the system who have been vaccinated up to the point in time t and have not left the compartment since the protective vaccination effect has fallen below a critical level. Vaccines experience waning immunity just like hosts recovered from natural infection, although disease-induced immunity generally induces a much longer lasting protection than vaccine-induced immunity [222, 225]. If the vaccine is successful, the host is immunized for some time. A waning effect can be included by introducing a rate $\mathcal{V}_2$, at which individuals transit from the compartment V back to S.

In Figure 4.3 a scenario with an applied all-or-nothing vaccine is illustrated. Compartments of unconfirmed cases, the class of susceptibles in quarantine, the hospitalized and the ICU compartment as well as deaths that are not related to SARS-CoV-2 are omitted here for transparency reasons. In the diagram, blue arrows from one to another compartment indicate a transition, whereas the compartment, from which a red dashed arrow originates, can infect susceptibles.

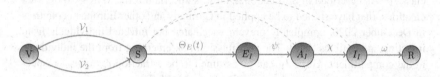

Figure 4.3 $SVE_IA_II_IR$ model with an all-or-nothing vaccine applied to the compartment S

The rate $\mathcal{V}_2$ can be modelled as a random variable that is distributed according to a probability distribution, which realistically reflects the waning protective effect of the respective vaccination. If leakiness is assumed, all vaccinees (vaccinated individuals) have a reduced probability of contracting the infection compared with the susceptibles in the compartment S. Thus a transition from the compartment S to a latently infected compartment E_V has to be incorporated into the model. A distinction has to be made between the vaccine efficacy, which indicates how effective the vaccine is under ideal conditions and a complete vaccine uptake, and the vaccine effectiveness, which measures the performance of the vaccine used in everyday circumstances [230].

The attack rates $A(t)$ for unvaccinated and $A_V(t)$ for vaccinated individuals in the model read [231]

$$A(t) = \frac{E(t) + E_I(t)}{S(t)} \quad \text{or} \quad A_V(t) = \frac{E_V(t)}{S(t)} \text{ , respectively}$$

and the vaccine efficacy can be defined as [232]

$$v_{es} = \frac{A(t) - A_V(t)}{A(t)} .$$

In other words, the parameter v_{es} is the efficacy of the leaky vaccine to reduce the contagiousness of the people transiting to the compartment E_V. It can also be described as the percentage reduction of contagion in a vaccinated group of people compared to an unvaccinated group [232]. In the sequel, the compartments E and E_I are condensed as $\hat{E}$, A and A_I as $\hat{A}$ and I_U and I_I as $\hat{I}$. Subsequently, the rate at which vaccinees progress from S to E_V is assumed as $(1 - v_{es}) \cdot \Theta_{\hat{E}}(t)$ here, with $\Theta_{\hat{E}}(t)$ the rate of transition from the class S to $\hat{E}$. A leaky vaccine can be assumed to not provide a complete protection even right after vaccination.

The individuals in the compartment E_V transit to a vaccinated asymptomatic contagious compartment A_V and from there to a vaccinated symptomatic infectious class I_V. As described in Subsection 2.4.4 it is uncertain if the SARS-CoV-2 vaccinations, that have started to be applied in Germany and other European countries in December 2020, completely prevent vaccinated but infected individuals from infecting susceptibles [233]. The transmission risk emerging from the individuals in the compartment A_V or I_V can be assumed to be a fraction v_{ea} or v_{ei}, respectively, of the transmission risk originating from $\hat{A}$ or $\hat{I}$. The parameters v_{ea} and v_{ei} describe the efficacy of the leaky vaccine to decrease the infectiousness of the vaccinated individuals in the compartment A_V or I_V compared to $\hat{A}$ or $\hat{I}$, respectively [234]. Both of them could moreover be defined to be age-dependent if age groups are included in the model.

If no vaccine protection is given anymore for an individual in the leaky vaccine scenario due to a waning effect, and the infection status of this individuals is known, which is difficult to find out in reality, the person can be assumed to transit to the next reachable unvaccinated compartment. Let x or y symbolize the fraction of vaccinees whose vaccine protection levels have fallen below a determined critical level per unit of time. They are both difficult to accurately determine but could be modelled by probability distributions. Then let $(1 - x) \cdot \chi_V$ or $(1 - y) \cdot \psi_V$ be the rate of transition of leaky-vaccinees from E_V to A_V or A_V to I_V without an assumed waning effect, respectively. Thus leaky-vaccinees progress from the class E_V to $\hat{A}$ at a rate $x \cdot \chi_V$ or from the compartment A_V to $\hat{I}$ at a rate $y \cdot \chi_V$, respectively.

In the $MATLAB$ implementation of an age group model with involved vaccination in this thesis, an all-or-nothing vaccine scenario is used. This has got the following reasons:

Firstly, the current authorized vaccines described in Subsection 2.4.3 have a verified efficacy against contagion of up to 95 %, which is very high, such that the portion of inefficient vaccinations can be regarded as negligible in a compartment model based on a large population. Secondly, it is not yet completely clear in how far the accessible vaccines protect vaccinated people from spreading the infection and their effect against certain known aggressive mutations has not been fully proven, but has also not been officially disproved. In the case of a possible transmissibility emerging from vaccinated individuals who cannot get sick with COVID-19, a transmission from the compartment V to S (which would be depicted by a dashed arrow) could be added to the all-or-nothing model. Thirdly, no reliable data concerning the fraction of vaccinated people with incomplete protection from COVID-19 are currently available, and it would be a complex task to provide such information in a reliable way.

Figure 4.4 depicts a scenario with an applied leaky vaccine. The parameter ω_V symbolizes the recovery rate of vaccinated infectious individuals. Again, blue arrows from one to another compartment indicate a transition, whereas the compartment, from which a red dashed arrow originates, can infect susceptibles.

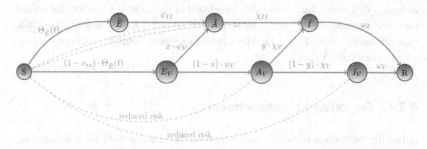

Figure 4.4 $S\hat{E}E_V\hat{A}A_V\hat{I}I_VR$ model with a leaky vaccine applied to the compartment S

It should be taken into account that vaccination strategies can be included in the model aside from the type of vaccination (leaky or all-or-nothing) and the efficacy of the regarded vaccine. Curiel and Ramírez [226] suggested 5 different strategies regarding a network-based population model, which can be combined with a compartment model. The first strategy prioritizes those individuals who have more contacts (more connected nodes in the network), the second one prioritizes the most or least controlling/influential individuals (nodes with a higher or lower betweenness), the third one vaccinates the oldest individuals in the population at first, and the last one randomly selects a portion of the population [226]. In the implementation in Chapter 6, a random selection of people per age group to be vaccinated is realized.

The *betweenness centrality* $b(v)$ of a node $v \in \bar{V}$ is defined as [227, p. 142]

$$b(v) = \sum_{i \neq j \neq u} \frac{\bar{\sigma}_{ij}(v)}{\bar{\sigma}_{ij}},$$

where $\bar{\sigma}_{ij}$ is the total number of shortest paths from node i to node j and $\bar{\sigma}_{ij}(v)$ is the number of those paths that pass through v. For each node in the graph, it is the number of these shortest paths that pass through the node. In contrast to that, the *degree centrality* $d(v)$ is the number of edges connected to a node v in an undirected graph i.e. equals the degree of a node [227, p. 121]:

$$d(v) = deg(v) .$$

Other centrality measures like the closeness centrality, the communicability centrality or the eigenvector centrality exist. In a network-based approach of epidemic modelling, a node could represent an individual in the population or an event, accidental meeting or planned meeting, which people participate in. Here, centrality measures of a node characterize the importance of the node in spreading the infection. For example, a superspreading event itself, or individuals who participate in a superspreading event, represented by a node in a graph would be distinguished by a higher centrality value.

4.2.12 Overview of Compartments

All of the mentioned transition parameters are listed along with their definitions, units and also selected fixed values or value bounds for the implementation in Chapter 6 of this thesis. The information if the respective parameter is fixed, estimated by algorithms or computed from other parameters is also included there.

Taking all of the mentioned compartments into account a table with the features of each class can be created.

Table 4.1. summarizes the defined characteristics of individuals of all classes in a compartment-wise manner. The leaky-vaccinated compartments E_V, A_V and I_V are not included here, but the all-or-nothing-vaccinated compartment V is incorporated.

The following possible further considerations with respect to the described compartments is emphasized:

All of the individuals in the compartments E_I, A_I and I_I are isolated, whereas those in the compartments E, A and I_U are not. It is reasonable to include a bureaucratic time delay between the receipt of a positive test result and the infliction of isolation in the model. The compartments E_I, A_I and I_I could be separated into a "confirmed but not yet isolated" and a "confirmed isolated" compartment each to account for the delay.

Table 4.1 Compartment Variables in the SARS-CoV-2-fitted SEIR model

Compartment Variable	Definition of comprised individuals
S	susceptible
S_q	susceptible, in preventive quarantine
V	vaccinated, not susceptible, not infected
E	(latently) infected, not infectious, asymptomatic (unconfirmed cases)
E_I	(latently) infected, not infectious, asymptomatic, in isolation (confirmed cases)
A	infected, infectious, asymptomatic (unconfirmed cases)
A_I	infected, infectious, asymptomatic, in isolation (confirmed cases)
I_U	infected, infectious, symptomatic (unconfirmed cases)
I_I	infected, infectious, symptomatic, in isolation (confirmed cases)
H	confirmed infected, infectious, in isolation, hospitalized, not in intensive care
C	hospitalized, in intensive care
R	recovered
D	deceased for disease-induced reasons

The case-fatality rate and the case-hospitalization rate could also be described by time-dependent functions and obtained via interpolation of reported case-fatality or -hospitalization data. Such data are uncertain due to probable misestimations of case-fatality and -hospitalization data. Multiple reasons for a misestimation of the CFR were explained in detail in Chapter 2. The CFR could be complemented by a turning point that depends on the ICU capacity present in the underlying population. This means that it could be assumed that the CFR punctually increases more rapidly when a certain ratio of ICU admissions is exceeded.

It is significant to note that quarantine and isolation measures, contact restrictions, social distancing and the realization of NPIs lead to remarkable social and economic costs, which should be reduced at a simultaneous containment of infection numbers. The connected optimization problem is not a part of the implementation of this thesis, but should be realized by means of numerical optimal control in future works.

4.3   Transmission in the SARS-CoV-2-fitted Model

In this section the rates of transition from the susceptible compartment to the exposed and quarantine compartments are analytically deduced.

In Subsection 4.3.1 SARS-CoV-2-suited variants of the transmission risk are mathematically explained.

Subsection 4.3.2 describes in which ways a contact, a quarantine as well as an isolation rate influence the transmission rate apart from the transmission risk. Certain time-dependent functions for the quarantine and contact rate are derived.

The purpose is to model the transmission, contact or quarantine rate in a way that fits the course of the respective rate that has been experienced in reality during the corona virus pandemic.

In Subsection 4.3.3, the compositions and definitions of the rates of the transitions emerging from the susceptible compartment are derived from the explanations of the previous two subsections.

The characteristics of a time delay model are given in Subsection 4.3.4 by making use of the transition rates explained in Subsection 4.3.3 to deduce the rates of transition from the exposed compartment to the compartments describing later states of infection.

4.3.1   Transmission Risk

Definition of the Transmission Risk
Let c_e be the average number of all contacts between a susceptible and an infectious person in the regarded population per unit of time. It is also called the *effective contact rate* of a population. Additionally, it can be asked to fulfil a certain condition like a short enough distance between the two involved people.

Let s be the average number of acquired secondary infections per unit of time in the regarded population.

Here, the *transmission risk* β concerning a specific infection and population is defined by the ratio of s to c_e. It is sometimes called the transmission probability and here mathematically defined as the ratio

$$\beta = \frac{s}{c_e}.$$

The effective contact rate c_e is smaller or equal to the *total contact rate* γ. The rate γ is defined by the average number of all contacts of one individual in the population per unit of time.

The product of the total contact rate γ and the transmission risk β is the rate at which susceptible individuals are infected per unit of time. The rate $\beta \cdot \gamma$ is used in Subsection 4.3.3. in order to define the rates of transmission in the SARS-CoV-2-fitted model.

The above definition of the transmission risk can also be called the *secondary attack rate*, which enables statements concerning the contagiosity of an agent. An Indian study verified that the secondary attack rate of 2019-nCoV estimated between December 2019 and June 2020 varies widely across countries with a lowest reported rate of 4.6 % and a highest of 49.56 % [228]. A meta-analysis of 54 studies with altogether 77,758 participants showed that the estimated overall household secondary attack rate of SARS-CoV-2 was 16.6 % and thus higher than observed secondary attack rates for SARS-CoV and MERS [229].

Mathematical Formulations of the Transmission Risk

The transmission risk is one of the parameters that are estimated in $MATLAB$ implementations performed in this thesis.

The transmission risk can be regarded as a constant value because for simplicity it can be assumed that the infection risk does not change over time. Nevertheless, it is discussed whether the wearing of face masks, the filter efficiency of the worn masks, the adoption of the temperature and shock ventilation in closed rooms reduce the infection risk [235].

All of these factors support the view that firstly, the infection risk depends on the personal realization of protection measures of every single person. In order to incorporate this into the model the susceptible compartment has to be separated in two or more classes, such that the individuals of each susceptible subclass are assumed to take protection and hygiene measure influencing the transmission risk seriously up to only a certain degree. Secondly, the mentioned factors sustain the view that the transmission risk is time-dependent due to the fact that protective measures like the expansion of compulsory mask wearing to more locations are taken at certain time instants and over specific time periods during the pandemic, such that the risk of transmission is reduced within the whole population during these periods.

Consequently, the transmission risk can be described by a function of time. The following function can be utilized as a transmission risk function to model the SARS-CoV-2-specific transmission rate:

$$\beta_1(t) = \begin{cases} \beta_0 & , t < \tau_m \\ \beta_0 \cdot e^{-j \cdot (t - \tau_m)} & , t \geq \tau_m \end{cases}$$

The function $\beta_1(t)$ decreases exponentially with respect to the time t. Here, β_0 is the transmission risk at the initial time $t_0 = 0$. The parameter $\tau_m \geq 0$ is the time instant at which the respective measure like compulsory mask wearing is taken and j is a constant value to control the transition rate [236]. The larger the control parameter j the steeper is $\beta_1(t)$. The function meets the value β_0 on the y-axis where $x = \tau_m$.

Moreover, the transmission risk related to SARS-CoV-2 seems to underlie seasonal fluctuations. Experimental data suggest that SARS-CoV-2 persistence on surfaces or in the air is sensitive to temperature, humidity as well as ultraviolet light. Furthermore, other environmentally sensitive respiratory viruses are more prevalent in winter. Therefore, climatic effects could be responsible for protective effects in dry and warm places and summer in general [237]. Low temperature and dry air impair and disrupt the integrity of the epithelial layer of the lungs, which might explain the winter seasonality of respiratory viruses [238]. Multiple studies showed that high temperature mitigated the transmission of the virus [238], but an analysis of transmission in 4 major provinces in Canada between January and May 2020 did not find a significant association between ambient temperature and transmission [239].

Studies have revealed that room or outside temperature and air humidity have an impact on the transmission rates and stability of respiratory viruses like SARS-CoV-2 [240, p. 83]. A mix of low humidity, temperature and sunlight may result in an impairment of the human local and systemic antiviral defence mechanisms, which result in an increased susceptibility to respiratory viruses in winter [240, p. 94].

This seasonality speaks for a time-dependent transmission risk function, which could be defined to be a positive continuous τ_P-periodic function $\beta_2(t)$, for which it holds

$$\beta_2(t + \tau_P) = \beta_2(t) .$$

Here, the parameter τ_P is a non-zero constant describing the length of the period after which an initial transmission risk is reached again. With regard to 2019-nCoV, a period of 6 months or $\tau_P = 26$ weeks is realistic because fluctuations in the transmission between summer and winter can be observed [237].

In the case of the involvement of seasonality and obtainment of the initial transmission probability value β_0 in the calendar week $t_0 = 10$ in winter, the time-dependent transmission risk can be defined as a sinusoidal function:

$$\beta_2(t) = \beta_0 \cdot \left(1 + \beta_P \cdot cos\left(\frac{2\pi t}{52}\right)\right)$$

The period of $\beta_2(t)$ is $\tau_P = \frac{2\pi}{52}$. The factor $\frac{1}{52}$ is selected since a year has 52 weeks. Consequently, the first local minimum is reached at $t = 26$, which is in the calendar week 36 in summer. The parameter β_0 is again the initial transmission risk, but also the mean transmission rate. The parameter β_P stretches the function along the y-axis and so defines the magnitude of the fluctuation. It has to be larger than zero in order that the function $\beta_2(t)$ does not get smaller than zero.

It should be added that stable oscillations like seasonal epidemic waves are fostered by some degree of immunity of the population [237].

Weather is a presumable factor to influence transmission of 2019-nCoV, but not at a scale sufficient to outbalance the effects of lockdowns or economic re-openings. So climate conditions are unlikely to inhibit SARS-CoV-2 prevalence in summer to a large measure. In addition, all pharmaceutical and non-pharmaceutical intervention measures are currently believed to affect the transmission of SARS-CoV-2 more than any environmental impacts. [237]. Consequently, the *MATLAB* implementations of Chapter 6 the transmission risk is regarded as a constant value. This is also reasoned by the fact that a seasonality already influences the transmission risk because β is estimated and adjusted based on the data of infected individuals reported from March 2020 until February 2021, which underlie seasonal fluctuations. More details concerning the parameter optimization are given in Chapter 6. Secondly, the obligation to wear face masks was not strictly locally extended or rescinded since the mask introduction in Germany in March 2020.

Only in December 2020, compulsory masks were extended to parking areas as well as city centres, and medical face masks became compulsory in early 2021. Thirdly, the obligation of shock ventilation was introduced in schools in autumn 2020, but the realization of this cannot be properly controlled.

Modifications of the Transmission Risk
It is significant to state that the transmission risk is not only infection- and intervention-specific but also depends on the contagious compartment containing individuals that infect susceptibles.

Hence, transmission probabilities proceeding from different compartments are distinct. In Section 4.2, 13 compartments were introduced, of which the compartments E_I, E, A_I, A, I_I, I_U, H and C are the infected ones. They are the *infected states* of the SARS-CoV-2-fitted model. Among these states, the 6 so-called *states-of-infectiousness* are given by the contagious compartments [195, p. 879] . Those are A, A_I, I_U, I_I, H and C in the SARS-CoV-2-fitted model.

According to Chapter 3 the number of individuals infected by a compartment $\mathcal{K}$ at time t is defined as

$$\beta \cdot \gamma(t) \cdot S(t) \cdot \mathcal{K}(t) .$$

There are six distinct infectious compartments in the SARS-CoV-2-fitted model. The differences between the transmission risks originating from the different classes $\mathcal{K} \in \{A, A_I, I_U, I_I, H, C\}$ have to be expressed. Subsequently, a factor $\varepsilon_{\mathcal{K}}$ is multiplied by β in the above term to obtain the specific transmission risk emerging from the class $\mathcal{K}$.

Modification factors concerning a transmission based on a contact of a susceptible with a member of the compartment $\mathcal{K}$ are described by the variable

$$\varepsilon_{\mathcal{K}} \in [0, 1] , \mathcal{K} \in \{A, A_I, I_U, I_I, H, C\} .$$

All individuals in the classes E and E_I are latently infected and not infectious. Those people who are isolated at home (classes I_I and A_I) can infect family members, people supplying them or mobile nursing services under certain conditions. Hospitalized individuals (classes H and C) can be assumed to be able to infect medical staff or frequent visitors. but nobody else.

If above-average hygienic conditions, visiting regulations and distancing and isolation regulations in hospitals are presupposed, it can however be assumed that all patients in the compartments H and C, comprising individuals in hospitals, do not have the ability to infect any susceptibles. Then it would hold that $\varepsilon_H = \varepsilon_C = 0$.

Certainly, a differentiation between the compartments H and C can take place by introducing a factor that reduces the transmission risk originating from one of the two classes compared to the probability emerging from the other one slightly to sharply.

The class I_U is the compartment that puts susceptibles at the highest infection risk during contact because the individuals in I_U are infected but undetected. A compartment which puts susceptibles at a higher risk of contagion during contact than another class is interpreted as a compartment with a higher transmission risk emerging from it.

It is clear that unconfirmed symptomatic infectious individuals spread the disease in an uncontrolled manner, particularly if no state measures are taken. Thus it is defined that $\varepsilon_{I_U} = 1$. The class I_U is assumed to put contacted susceptibles at a higher infection risk than the isolated symptomatic individuals in the compartment I_I. The same holds for the contagiousness relation between the classes A_I and A. Additionally, it is assumed that unconfirmed asymptomatic cases (A) have a higher risk of infecting susceptibles than confirmed symptomatic individuals (I_I), since they are often not even aware of being infectious. The individuals in the class I_I are assumed to have a higher risk of infecting susceptibles than hospitalized patients as all hygiene and distancing rules are assumed to be taken more serious by hospital personnel than at home.

By implication, the following inequalities between the modification factors can be inferred:

$$1 = \varepsilon_{I_U} > \varepsilon_A > \varepsilon_{I_I} > \varepsilon_{A_I} > \varepsilon_H > \varepsilon_C .$$

The average length of the period of contagiousness of a certain group of any infected individuals determines how long the individuals in this group can infect susceptible individuals.

Let the time period until loss of infectiousness be $P_{\mathcal{K}_j}$ for a compartment $\mathcal{K}_j$, $j = 1, 2, \dots$.

Then $P_{\mathcal{K}_j}$ can be normalized by the average length of stay $\mathcal{D}_{\mathcal{K}_j}$ in the respective class to obtain a ratio $Z_{\mathcal{K}_j}$ of the infectious period to the average length of stay in the respective compartment

$$Z_{\mathcal{K}_j} = \frac{P_{\mathcal{K}_j}}{\mathcal{D}_{\mathcal{K}_j}}, \ j = 1, 2, \dots .$$

Let $Z_{\mathcal{K}_1}$ be smaller than $Z_{\mathcal{K}_2}$ for two arbitrary classes $\mathcal{K}_1, \mathcal{K}_2$. For instance, it can be defined that $\varepsilon_{\mathcal{K}_2} = \frac{\varepsilon_{\mathcal{K}_1} \cdot Z_{\mathcal{K}_2}}{Z_{\mathcal{K}_1}}$. Subsequently, the modification factor $\varepsilon_{\mathcal{K}_2}$ is defined as larger than $\varepsilon_{\mathcal{K}_1}$. In this case the normalized length of the period until loss of contagiousness is compared between two compartments and incorporated into the respective modification factor computations in order to be able to reduce the transmission risk of the one class compared to the other class in the correspondent model.

4.3.2 Contact, Quarantine and Isolation Rates

In the explained model, a distinction between a transition from the class S to class S_q, E or E_I has to be made. Susceptibles are assumed to be put in quarantine and transit to the compartment S_q according to the quarantine rate q. A fraction $1 - q$ is not put in quarantine and exposed to the infection risk. It is still assumed that the probability of an infection by contact is β. Consequently, the probability of not contracting an infection based on a contact with an infected person is $1 - \beta$. A portion $\mathcal{I}$ of all susceptibles who have not been put in quarantine and contract the infection is tested and isolated. This portion progresses from the class S to E_I.

A portion $1 - \mathcal{I}$ of all the susceptibles who are not in quarantine and have just got infected is not tested or isolated. This fraction of infected individuals is unconfirmed cases and transits to the compartment E_I.

The reason for the following choice of a time-dependent contact rate function is the fact that in the course of the spread of the pandemic contacts are increasingly restricted in the public as well as in private and contact rates decrease due to an increasing number of suspected cases and more cautiousness of health offices and society in general.

Various scientific papers that deal with modelling the SARS-CoV-2-pandemic estimate the contact rate and, if included, the quarantine rate as constant values. To express an increase in intervention and restriction measures in terms of a selected contact rate, a function behaving exponentially with respect to the time $t \in [0, T]$ can be used [241].

Let the contact rate function be an exponential function, γ_0^{exp} the initial contact rate and γ_{min}^{exp} the minimum contact ratio under the applied specific control strategies. Then the contact rate function can be defined as

$$\gamma^{exp}(t) := (\gamma_0^{exp} - \gamma_{min}^{exp}) \cdot e^{-d_1 \cdot t} + \gamma_{min}^{exp},$$

where $(\gamma_0^{exp} - \gamma_{min}^{exp}) + \gamma_{min}^{exp} = c_0^{exp} = \gamma^{exp}(0)$ denotes the y-intercept of the function, which is the initial contact rate. Then it holds that

$$\lim_{t \to \infty} \gamma^{exp}(t) = \gamma_{min}^{exp} \text{ with } 0 < \gamma_{min}^{exp} < \gamma_0^{exp}.$$

If the parameter d_1 lies in the interval $(0, 1)$, the function is flatter than it was if d_1 was set to the value 1, such that the function is stretched. If d_1 is larger than 1, the function is tighter than it was if d_1 was set to the value 1, such that the function is compressed. Moreover, the smaller $d_1 > 1$ or the larger γ_{min}^{exp} or the smaller γ_0^{exp} is, assuming that the other respective parameters are invariable then, the more compressed is $\gamma^{exp}(t)$.

Additionally, let q be the quarantine rate, meaning the rate at which individuals are placed in quarantine (which means in the compartment S_q). It is defined as

$$q = \frac{\text{number of individuals placed in quarantine per unit of time}}{\text{number of all susceptibles in the population per unit of time}}.$$

Let $\mathcal{I}$ be the isolation rate, which is the rate at which individuals are positively tested and in this way isolated. It is defined as

$$\mathcal{I} = \frac{\text{number of individuals who are positively tested and isolated per unit of time}}{\text{number of all susceptibles in the population per unit of time}}.$$

Analogous to the function $\gamma^{exp}(t)$, the quarantine rate q can be described by an exponential function $q^{exp}(t)$.

Therefore, let the value q_{max}^{exp} be the maximum contact rate under the applied intervention measures. It is the limit of the quarantine rate function $q^{exp}(t)$. It holds that

$$\lim_{t \to \infty} q^{exp}(t) = q_{max}^{exp} \text{ with } 0 < q_0^{exp} < q_{max}^{exp}$$

with q_0^{exp} the initial quarantine rate in the absence of any control strategies. The quarantine rate can then be described by the time-dependent function

$$q^{exp}(t) := (q_0^{exp} - q_{max}^{exp}) \cdot e^{-d_2 \cdot t} + q_{max}^{exp},$$

where $(q_0^{exp} - q_{max}^{exp}) + q_{max}^{exp} = q_0^{exp} = q^{exp}(0)$ denotes the y-intercept of the function, which is the initial quarantine rate.

In the sequel, the term contact ratio (quarantine ratio) is used for the value of a contact rate (quarantine rate) at a specific point in time t. In Figure 4.5 the contact and quarantine rate functions for the values $\gamma_0^{exp} = 80$, $q_0^{exp} = 0$, $\gamma_{min}^{exp} = 28$, $q_{max}^{exp} = 0.3$, $d_1 = d_2 = 0.05$ are illustrated.

The unit of the contact ratio on the x-axis is $\frac{contacts}{week}$.

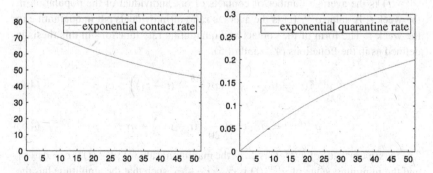

Figure 4.5 Exponentially decreasing contact rate with an initial 80 contacts per person per week and a limit of 28, and exponentially increasing quarantine rate with an initial value 0 and a limit of $\frac{3}{10}$ of the population, depicted in a progress over 52 weeks.

Figure 4.5 shows a contact rate that strives towards a value of 28 contacts per person per day as well as a quarantine rate striving against a value of 30 % of the population put in quarantine. The selection of the 2 rates as exponential functions is realistic for times in which the number of new infections and as a consequence the magnitude of intervention measures rises. After incidence rates sank all over the world in the summer of 2020, the numbers of new infections increased in various countries in the autumn of 2020. Quarantine rates increased and the average weekly number of contacts per person decreased. Since the data used for parameter optimization and prognoses in this thesis generally refer to pandemic times, in which the SARS-CoV-2 incidence increased or decreased all over the world, it seems móre reasonable to model $\gamma(t)$ ($q(t)$) as a function that reaches a maximal (minimal) value at the time when intervention measures reached their minimum in the summer and afterwards decreases (increases) again. Thus it is more realistic to model the contact and quarantine rate as periodic than exponential functions. Trigonometric cosine functions are selected here.

The course of the SARS-CoV-2 incidence in Germany, and similarly in Europe and other countries all over the world (cf. Chapter 2), implies that restrictive measures taken by the respective state increased until April 2020, were continuously reduced between May and September 2020 and were widely extended when the second wave of the pandemic announced itself in the beginning of October 2020.

The regarded initial point in time, which the initial contact rate γ_0^{trig} and quarantine rate q_0^{trig} refer to, is the beginning of the 10th calendar week in 2020. A value $\gamma^{trig}(t)$ is the average number of contacts of one individual of the population in the week t. A value $q^{trig}(t)$ is the average ratio of individuals put in quarantine in the week t. The form of the contact and quarantine rate functions in this thesis is defined as in the Equations (4.4) and (4.5).

$$\gamma^{trig}(t) = (c_2 - c_0) \cdot cos\left(\frac{\pi}{20} \cdot (t - z_1)\right) + c_1 \tag{4.4}$$

$$q^{trig}(t) = q_1 \cdot cos\left(\frac{\pi}{20} \cdot (t - z_2)\right) + q_1 . \tag{4.5}$$

The maximum value of $q^{trig}(t)$ is $2 \cdot q_1$, the maximum value of $\gamma^{trig}(t)$ is $c_2 - c_0 + c_1$ and the minimum value of $\gamma^{trig}(t)$ is $c_0 + c_1 - c_2$, such that the amplitude has the width $2 \cdot c_2 - 2 \cdot c_0$. The parameters z_1 and z_2 determine the shift of the respective function on the x-axis. Both functions are of the general form

$$a \cdot cos\left(b \cdot (t + w)\right) + d ,$$

where a symbolizes the width of the amplitude of the function, b is the length of the period, w represents the shift on the x-axis and d defines the shift on the y-axis in comparison to the function $cos(t)$.

Under the circumstances of an absent pandemic the average number of contacts per person per week can for instance be assumed to be larger than 100 (meaning more than an average 14 contacts of any kind per person per day) and the quarantine rate to be zero or very slightly larger than zero.

In Figure 4.6, the contact and quarantine rate functions for the values $c_0 = 10, c_1 = 70, c_2 = 30, q_1 = 0.002, z_1 = 35$ and $z_2 = 15$ are depicted. Thus the contact rate fluctuates within the interval [50,90], which shows that a scenario of certain precautions for the present COVID-19 pandemic is assumed in the two figures below.

The initial contact rate (quarantine rate) is assumed to already be lower (higher) than in conditions with an absent pandemic here because in the beginning of the outbreak in Germany in the end of February/beginning of March 2020, people got alarmed and first recommendations of not going out were given by the authorities. These recommendations were replaced by bans on going out in mid-March 2020 in several European countries.

The unit of the contact ratio on the x-axis in Figure 4.6 is $\frac{contacts}{week}$.

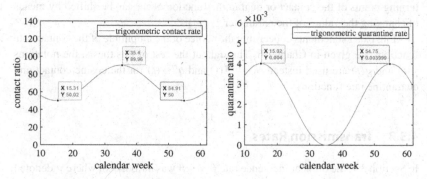

Figure 4.6 Periodic contact rate with an maximum number of 90 and a minimum of 50 contacts per person per week, and periodic quarantine rate with an initial value and maximum of 0.4 % of the population, depicted in a progress over 62 weeks.

In Figure 4.6, the contact rate (quarantine rate) is assumed to sink until a minimum value of 50 (minimum value of around 0) is attained in the 15th calendar week, which is in April. Then it rises until a maximum value of 90 (maximum value of 0.004) is reached in the 35th calendar week, which is in August 2020. At next, the respective periodic trigonometric function of the contact rate (quarantine rate) sinks again. The contact rate attains the same minimal value as in April in December. In this example, the quarantine rate is sinking (rising) when the contact rate reaches its minimum (maximum), and attains its minimal (maximal) value when the contact rate is increasing (decreasing).

The course of the proposed quarantine rate in Figure 4.6 resembles the course of the stringency index of Germany in the way that maximum stringency in terms of interventions was roughly present between the calendar weeks 13 and 18 as well as from the calendar week 51, and minimum stringency roughly between the calendar weeks 35 and 42. When comparing the progressions of the stringency index and the quarantine rate it has to be kept in mind that the stringency index is composed of 13 different restriction measures but quarantine is only one means of intervention. For instance, the population got alarmed and first state interventions like shop closures came into effect before quarantine measures were systematically implemented. Moreover, a bureaucratic time lag always has to be respected in the consideration of the quarantine rate. This kind of delay can be included to express the fact that quarantine or contact restrictions are often imposed later than would be necessary to prevent secondary infections, and be able to distinguish between the time lags of contact restriction and quarantine measures. The times t at which the turning points of the contact or quarantine function occur can be shifted by means of changing the values of the parameter z_1 (z_2).

More exact statements concerning the choice of the amplitude of the contact rate functions are given in Chapter 6. Throughout the rest of this thesis, the notations $\gamma(t)$ and $q(t)$ are used instead of $\gamma^{trig}(t)$ and $q^{trig}(t)$ for the cosine contact and quarantine rate functions.

4.3.3 Transmission Rates

In Section 3.1, the bilinear incidence rate $\tilde{\beta} = \gamma\beta$ was introduced, where γ denoted a constant contact rate and β denoted a constant probability of transmission per contact, such that $\tilde{\beta} \cdot S \cdot I$ would be the number of individuals who become infected per unit of time if one single infectious compartment I was assumed.

In the proposed SARS-CoV-2-fitted model, the contact rate is assumed to be a cosine function as suggested in the previous subsection. Thus $\gamma(t)$ is the average number of contacts of all individuals in the system at time t, which means in the week t. Contact here comprises contact with any other people in the regarded population.

State interventions like prohibitions to enter public places, restraining orders or bans on going out can be incorporated into the model by decreasing the contact rate $\gamma(t)$.

Proceeding from Chapter 3, omitting the time reference t and not regarding quarantine for the first, the number of individuals infected by people in the compartment I_U can be defined as

$$I_{new}^{I_U} := \beta \cdot \gamma \cdot \frac{S}{N} \cdot I_U .$$

Resorting to the remarks of Subsection 4.3.1, the number of individuals who get infected by a class $\mathcal{K} \in \{A,\ A_I,\ I_I,\ H,\ C\}$ per unit of time is

$$I_{new}^{\mathcal{K}} := \beta \cdot \gamma \cdot \frac{S}{N} \cdot \varepsilon_{\mathcal{K}} \cdot \mathcal{K} .$$

By adding the equations $I_{new}^{\mathcal{K}}(t)$ for all compartments $\mathcal{K}$, it can be seen that the number of susceptibles infected by all infectious classes at time t is given by

$$I_{new} := \beta \cdot \gamma \cdot \frac{S}{N} \cdot \tilde{X} \text{ with } \tilde{X} := I_U + \varepsilon_A \cdot A + \varepsilon_{A_I} \cdot A_I + \varepsilon_{I_I} \cdot I_I + \varepsilon_H \cdot H + \varepsilon_C \cdot C , \quad (4.6)$$

which represents a bilinear incidence. Incorporating a quarantine rate $q(t)$ and a case-confirmation/isolation rate $\mathcal{I}$ into (4.6), the different rates of transition from the class S to S_q, E or E_I, respectively, can be derived:

$$\theta_{S_q}(t) = \left(1 - \beta\right) \cdot \gamma(t) \cdot q(t) , \quad (4.7)$$

$$\theta_E(t) = \beta \cdot \gamma(t) \cdot \left(1 - q(t)\right) \cdot \left(1 - \mathcal{I}(t)\right) , \quad (4.8)$$

$$\theta_{E_I}(t) = \beta \cdot \gamma(t) \cdot \left(1 - q(t)\right) \cdot \mathcal{I} . \quad (4.9)$$

The transition rates given in the Equations (4.7) to (4.9) can be called transmission rates because of their dependence on the transmission risk β. The number of individuals moving from the compartment S to S_q or E or E_I at time t can be described by $\Theta_{S_q}(t) \cdot S(t)$, $\Theta_E(t) \cdot S(t)$ or $\Theta_{E_I}(t) \cdot S(t)$, respectively, for which holds:

$$\Theta_{S_q}(t) \cdot S(t) := \theta_{S_q}(t) \cdot \tilde{X}(t) \cdot S(t) = \left(1 - \beta\right) \cdot \gamma(t) \cdot q(t) \cdot \tilde{X}(t) \cdot S(t) , \qquad (4.10)$$

$$\Theta_E(t) \cdot S(t) := \theta_E(t) \cdot \tilde{X}(t) \cdot S(t) = \beta \cdot \gamma(t) \cdot \left(1 - q(t)\right) \cdot \left(1 - \mathcal{I}\right) \cdot \tilde{X}(t) \cdot S(t) , \qquad (4.11)$$

$$\Theta_{E_I}(t) \cdot S(t) := \theta_{E_I}(t) \cdot \tilde{X}(t) \cdot S(t) = \beta \cdot \gamma(t) \cdot \left(1 - q(t)\right) \cdot \mathcal{I} \cdot \tilde{X}(t) \cdot S(t) . \qquad (4.12)$$

In other words, the function $\Theta_K(t)$ defines the time-dependent rate of transition from the compartment S to $\mathcal{K}$, $\mathcal{K} \in [S_q, E, E_I]$.

It should be noted that according to the transmission rates in the Equations (4.7) to (4.9), people in the compartment S_q are assumed not to be infected. **If a susceptible catches the infection, it is not put in quarantine, but isolated if its infection is detected.**

An alteration of the parameter β, the parameter ε_K for different compartments $\mathcal{K}$, or the rates $q(t)$ and $\gamma(t)$ has a distinct impact on the model and compartment size predictions made on the basis of the model.

Relevant causes of the modification of these parameters and rates are the following ones:

The transmission risk β is enlarged if the portion of mutations with a higher transmissibility than the initially known novel corona virus among all infections per unit of time becomes larger. It is clear that the amount of easier transmissible mutated versions of the virus among all infections has to be taken into account to determine the increase in β. Mutations can also lead to larger ε_K for $\mathcal{K} \in [A_U, A_I, I_U, I_I, H, C]$ if people in a certain infected compartment transmit the mutated virus relatively more often than people in other infected compartments compared to the original version of the virus. For instance, symptomatic people might transmit a mutant form of the virus even faster than the original version of SARS-CoV-2, but asymptomatic individuals might not. This could also apply to specific age groups.

In general, leaky-vaccinees are less probable to transmit the viral infection compared to unvaccinated people. The introduction of a well-organized population-wide vaccination program using effective leaky vaccines as presented in Figure 4.4 leads to a reduced growth of the compartments E_I, A_I and I_I over the course of time. The transmission risk emerging from the leaky-vaccinated individuals in the compartments A_V and I_V is also smaller than the transmission risk originating from the people in the classes A_I and I_I. This can be characterized by transmission coefficients $\varepsilon_{A_V} = v_{ea} \cdot \varepsilon_{A_I}$ and $\varepsilon_{I_V} = v_{ei} \cdot \varepsilon_{I_I}$, with $v_{ea} < 1$, $v_{ei} < 1$.

Another case is regarding a model without a vaccinated compartment, although a leaky vaccination of population members is principally possible in reality as in the $SIHCDR$ model applied to German and Swedish COVID-19 data in Chapter 6. If it is assumed that susceptible people can be vaccinated, the transmission risk β

emerging from vaccinees in a later infected state i.e. from a portion of the infected compartment is reduced. It is certainly more transparent to directly include leaky-vaccinated compartments in the model, but this may not be possible due to missing reported data. Moreover, a vaccination program with leaky vaccines can reduce the transmission modification factors $\varepsilon_{\mathcal{K}}$ for $\mathcal{K} \in [I, H, C]$ relatively to each other. A more extensive vaccination program indicates a smaller transmission rate, whereas a missing or badly organized vaccination program with few accessible vaccines implies a comparatively larger transmission rate.

If an all-or-nothing vaccine scenario as depicted in Figure 4.5 is given, vaccinees are assumed not to become sick with the disease or be able to transmit the infectious agent as a presupposition, at least until a possible waning effect occurs.

It has to be stressed that an extension and better organization of testing programs leads to more case-confirmations and isolations, that would indirectly affect the amount of occurring transmissions. Moreover, this would increase the quarantine rate since more contact persons of infected people would be put in quarantine. Stricter and controlled isolations of the people in a compartment $\mathcal{K}$ in a regarded system might result in a decrease in the size of $\varepsilon_{\mathcal{K}}$, $\mathcal{K} \in [A_I, I_I, H, C]$ compared to lightly treated isolation measures.

Moreover, realized intervention measures such as lockdowns and the time periods over which those measures are implemented have large effects on the rates $q(t)$ and $\gamma(t)$. Increased state interference and precautious measures should usually lead to an increase in the parameter q_1 in the rate $q(t)$ and a decrease in the parameter c_1 in the rate $\gamma(t)$. The effects are detailed in Section 6.3 in Chapter 6.

4.3.4 Time Delay

We let the average constant length of the latency period be $\mathcal{L}$ and the average constant length of the period between the acquisition of infectiousness and symptoms be τ. Moreover, we let the rates in the Equations (4.7) to (4.9) describe the transitions from the compartment S to S_q, E, E_I.

Apart from a bureaucratic time lag mentioned in Subsection 4.2.12, a time delay expressing how the latency period and the incubation period influence the transition from the exposed to asymptomatic infectious or asymptomatic infectious to symptomatic infectious class can be involved in the model on the basis of the transmission rates. In the SARS-CoV-2-fitted model, all individuals who progress from the susceptible to the exposed class will also move to further infected compartments unless they die, and the need for specific infected compartments whose size is difficult to estimate can be eliminated by introducing time delay in the epidemic model with the aid of DDEs. DDEs are also applied in other scientific areas such as biology, chemistry, physics and mechanics.

A *delay differential equation* (DDE) for $z(t) \in \mathbb{R}^n$ is defined as

$$\frac{dz(t)}{dt} = g\left(t, z(t), z_t\right),$$

where $g : \mathbb{R} \times \mathbb{R}^n \times C^1(\mathbb{R}, \mathbb{R}^n) \rightarrow \mathbb{R}^n$ is a function, and $z_t = \left\{z(t_{past})|t_{past} \leq t\right\}$ is the trajectory of the solution in the past [242].

The delay can be pointwise i.e. discrete as

$$\frac{dz(t)}{dt} = g\left(t, z(t), z(t - t_{past\,1}), z(t - t_{past\,2}), ..., z(t - t_{past\,M})\right)$$

with $t_{past\,1} > ... > t_{past\,M}$, or continuous as

$\frac{dz(t)}{dt} = g\left(t, z(t), \int_{-\infty}^{0} z(t - t_{past}) d\hat{\leq}(t_{past})\right).$

A form of DDE that is often applied in medical and biological modelling is the delayed recruitment equation, which is given in Equation (4.13).

$$\epsilon \cdot y' = -y + g\left(\hat{\lambda}, y(t - 1)\right), \tag{4.13}$$

where $\epsilon = \frac{t_{past}}{t}$ is the linear decay time of the dependent variable, and $g(\hat{\lambda}, y)$ is a nonlinear function depending on t and a control parameter $\hat{\lambda}$ [243, p. 24].

It should be noted that the time-dependent solution of a DDE is not uniquely determined by its initial state at a single point in time. Instead, the solution profile on an interval with a length equal to the delay is needed. An infinite-dimensional set of initial conditions between $t = -t_{past}$ and $t_0 = 0$ is necessary such that DDEs are infinite-dimensional problems [243, p. 2]. Beyond that, a DDE has to be equipped with the solution at the initial point at times prior to this initial point [243, p. 3]. Furthermore, oscillatory instabilities are a point for discussion with regard to DDEs. Often addressed issues are first-order nonlinear DDEs exhibiting square-wave oscillations and second-order nearly conservative equations exhibiting both periodic and quasi-periodic oscillations [243, p. 23].

Making use of the average sojourn time in the respective infected compartment, the rate of transition from the class A_I to I_I could be described by the equation

$$\chi_{II} = \frac{1 - \mu \cdot \tau}{\tau}$$

and the rate of transition from the compartment E_I to A_I can be given by

$$\psi_{II} = \frac{1 - \mu \cdot \mathcal{L}}{\mathcal{L}}$$

because all individuals (except for deceased individuals) in the compartment A_I (E_I) are assumed to have transited to the class I_I (A_I) as soon as the period τ $(\mathcal{L})$ has passed.

In the sequel, we regard a system experiencing the dynamics connected to the outbreak of COVID-19 expressed through the SARS-CoV-2-fitted model.

Let ζ_I be the proportion of exposed (not isolated) individuals who develop confirmed infectiousness and $\zeta_U = 1 - \zeta_I$ be the proportion of exposed (not isolated) ones who develop infectiousness that is not detected and confirmed. Let κ_I be the proportion of asymptomatic (not isolated) individuals who develop confirmed symptoms and let $\kappa_U = 1 - \kappa_I$ be the proportion of asymptomatic (not isolated) people who develop symptoms that are not detected and confirmed. Let the probability that an individual survives the latent period $[t - \mathcal{L}, t]$ be $e^{-\mu \cdot \mathcal{L}}$ and the probability that a person survives the incubation period $[t - T^I, t]$ be $e^{-\mu \cdot T^I}$ [244, p. 121]. Then the rate at which individuals transit from the class S to the compartment A_I in the compartment model with time delay is given by the function

$$\upsilon_{II}(t) = \Theta_{E_I}(t - \mathcal{L}) \cdot e^{-\mu \cdot \mathcal{L}} \tag{4.14}$$

and the rate at which individuals transit from the class S to the compartment A or A_I is given by

$$\upsilon_U(t) = \Theta_E(t - \mathcal{L}) \cdot \zeta_U \cdot e^{-\mu \cdot \mathcal{L}} \tag{4.15}$$

or

$$\upsilon_I(t) = \Theta_E(t - \mathcal{L}) \cdot \zeta_I \cdot e^{-\mu \cdot \mathcal{L}} \text{ , respectively.} \tag{4.16}$$

Analogously, the rate at which individuals progress from the class S to the compartment I_I is given by

$$\varrho_{II}(t) = \Theta_{E_I}(t - \mathcal{L} - \tau) \cdot e^{-\mu \cdot T^I} = \Theta_E(t - T^I) \cdot e^{-\mu \cdot T^I} \tag{4.17}$$

and the rate at which individuals transit from the class S to the compartment I_U or I_I is given by

$$\varrho_U(t) = \Theta_E(t - \mathcal{L} - \tau) \cdot \kappa_U \cdot e^{-\mu \cdot T^I} = \Theta_E(t - T^I) \cdot \kappa_U \cdot e^{-\mu \cdot T^I} \tag{4.18}$$

or

$$\varrho_I(t) = \Theta_E(t-\mathcal{L}-\tau)\cdot\kappa_I\cdot e^{-\mu\cdot\mathcal{T}^I} = \Theta_E(t-\mathcal{T}^I)\cdot\kappa_I\cdot e^{-\mu\cdot\mathcal{T}^I} \text{ , respectively.}\quad (4.19)$$

If the time-dependent transition rates in the Equations (4.17) to (4.19) are used, the exposed and asymptomatic infectious compartments do not have to be stated in the model since they are indirectly included via the length $\mathcal{T}^I$ of the incubation period. Compartment models resulting from the inclusion of transition rates that contain a certain latent or incubation delay belong to the class of DDEs.

The rate $\hat{\phi}$ for the transition from the compartment S_q to S could be replaced by the rate $\Theta_{S_q}(t-14)$ as well.

Model Speci cations

5

This chapter presents specifications of the introduced SARS-CoV-2-fitted model in the form of generally possible model variants, the provision of adequate systems of ODEs, model diagrams for two model variants and the SARS-CoV-2-fitted model itself, as well as the introduction of so-called *reproduction numbers* and the computation of the *basic reproduction number* based on the systems of ODEs and previously explained transmission dynamics.

In Section 5.1, distinct variations of the consideration of all compartments presented in Section 4.2 are explained. Models that are reduced in the number of compartments compared to the SARS-CoV-2-fitted model emerge from this. They are created substantially for the purpose of providing models that are well adapted for an implementation in $MATLAB$ in Chapter 6.

The developed framework consists among others of a system of ODEs that describes mathematically the dyna- mics of the SARS-CoV-2-fitted model. A definition of the connected initial value problem is given in Subsection 5.2.1. Two models with reduced numbers of compartments are presented in Subsection 5.2.2 along with their corresponding systems of ODEs. A system of ODEs for the enhanced model presented in Section 4.2 is derived in Subsection 5.2.3.

A model variant reduced in the number of compartments characterized by specific features but consisting of 3 different age groups (*age group model*) is introduced in Subsection 5.2.4 in the same way. The form of the incorporated transmission rate is derived as well.

In Section 5.3 the basic, control and effective reproduction numbers are defined. The principle of computing the basic reproduction number based on the SARS-CoV-2-fitted model is mathematically derived.

© The Author(s), under exclusive license to Springer Fachmedien Wiesbaden GmbH, part of Springer Nature 2021
S. M. Treibert, *Mathematical Modelling and Nonstandard Schemes for the Corona Virus Pandemic*, BestMasters, https://doi.org/10.1007/978-3-658-35932-4_5

5.1   Model Variants

Aside from using the SARS-CoV-2-fitted model, simplified sub-models can be created for further investigation and analysis, where certain classes are omitted or pooled for various purposes. The exclusion or omission of a compartment of people in quarantine, the pooling of isolated and not isolated compartments that are otherwise characterized by the same features, the exclusion of unconfirmed cases as well as a way of pooling all infected compartments are presented in this section.

5.1.1   Exclusion of Quarantine or the Quarantine Compartment

The first possibility of creating a condensed model is to disregard any means of quarantine. The target is to model a scenario in which no imposed quarantine restrictions the are present. To realize this, the compartment S_q and quarantine rate $q(t)$ are omitted, such that no susceptible is protected from the infection risk for a quarantine period. All of the further mentioned pooling methods can be performed in addition to the nonobservance of quarantine.

In the later implementations of different model variants, reported data of compartment sizes that mirror the exact numbers of individuals in the compartments as reliably as possible are needed. The target of the implementations is minimizing the error between the reported weekwise data and model-generated data of compartment sizes by adapting certain estimated model parameters. The used minimization techniques as well as details concerning the necessity of reliable compartment size data for the implementations are explained in Chapter 6. It is almost impossible to find reliable weekwise data concerning the exact number of individuals who are put in precautious quarantine or decide for inflicting self-quarantine. It is also difficult to accurately assess how many individuals are released from quarantine per unit of time.

For this reason, the compartment S_q can be omitted, too. To not ignore quarantine when leaving out the compartment S_q, a quarantine rate $q(t)$ can still be used in the rate of transition from the susceptible to the exposed class by multiplying the transmission risk β by the factor $(1 - q(t))$.

As regards content, this means that a proportion $q(t)$ of the population cannot get infected at time t since it is assumed to be in quarantine. NPIs can be regarded as influencing the rate $q(t)$ if restrictions other than sole quarantine imposed on the population are taken into account.

It does not mean that a fixed proportion $q(t)$ is excluded from susceptibility forever. This would be the case if individuals reached the compartment S_q at a rate θ_{S_q} but did not return to the class S. The impact of $q(t)$ on the transmission rate was explained in Section 4.3. The parameters in the quarantine rate or a constant quarantine ratio are estimated in the optimization part of the implementation. Clearly, distinct quarantine and state intervention scenarios have to be defined and associated with certain values assigned to the parameters occurring in the quarantine rate.

5.1.2 Pooling of Isolated and not Isolated Compartments

For the purpose of decreasing the number of compartments in order to obtain a smaller model, the classes E and E_I can be combined as one latently infected class $\hat{E}$, which includes exposed individuals regardless of their isolation status. As a result, it makes sense to also combine the compartments A and A_I to a class $\hat{A}$, and the classes I_I and I_U to a compartment $\hat{I}$ to generally not distinguish between isolated and not isolated individuals. **The rate of transition from the compartment S to $\hat{E}$ is independent of an isolation rate $\mathcal{I}$ then.** Thus not all of the individuals in the class $\hat{E}$ are regarded as isolated and so not (positively) tested, which makes the search for compartment size data more difficult.

5.1.3 Exclusion of Unconfirmed Infected Cases

The nonobservance of isolation in the model reduces the number of equations in the associated system of ODEs and omits the isolation rate, but does not simplify the search for compartment size data as explained in the previous subsection. A different method of decreasing the number of model compartments is the omission of the classes E, A and I_U. At each instant of time t, $E(t)$, $A(t)$ and $I_U(t)$ represent all of the infected individuals who have not been confirmed up to the time instant t. Since confirmation of a case entails isolation, E, A and I_U also symbolize cases that are not isolated. They represent the dark figure of the disease. If these compartments are not incorporated into the model, unconfirmed infected cases are excluded from the model.

The simplification of a model in which the unconfirmed cases are omitted compared with models in which the dark figure is included is accounted for by the non-existence of exact data concerning the size of the dark figure of the corona virus pandemic.

5.1.4   Pooling of Infected Compartments

The difficulty of using a model with a separate exposed, asymptomatic infectious and symptomatic compartment is to find reliable data concerning the numbers of symptomatic infectious, asymptomatic infectious as well as asymptomatic latently infected individuals per country and per time instant t. In most reliable data sources a distinction in infected cases is made at most between symptomatic and asymptomatic confirmed cases.

Due to missing data the separation of latently infected and asymptomatic infectious individuals, who both do not show any symptoms but are infected, into different classes in a model is substantially difficult to implement.

This complicity is overcome by pooling the compartments E_I and A_I to a single class $\tilde{A}_I$ and the compartments E and A to a single class $\tilde{A}$, that together comprise all asymptomatic infected individuals disregarding their status of contagiousness.

Following the remarks in Chapter 2 and Section 4.2, infected cases that are asymptomatic infectious or not even yet contagious are not detected often but rather by chance, which for instance means when they are tested because a family member or close friend is a confirmed infected case or they have returned from a high risk area. Hence, the sequence of transitions between the susceptible and infectious stage that occurs most often is $S \rightarrow E \rightarrow A$. It is logical that an infection with 2019-nCoV is discovered and reported most often when individuals have just developed symptoms. As the transition $A \rightarrow I_I$ is more probable than $S \rightarrow E_I$ or $E \rightarrow A_I$ and I_I comprises confirmed symptomatic cases, it makes sense to include the compartment I_I into a model. For an additional model reduction, the compartment $\tilde{A}_I$ can be bundled with the compartment I_I to obtain a single confirmed infected compartment I. In this case not even a distinction between asymptomatic and symptomatic cases is necessary with respect to data search. For data search concerning the transition rate from I to R the time from case-confirmation to recovery is relevant.

If all confirmed infected classes are pooled, solely the number of new infections per time instant t has to be found to provide an implemented algorithm with the necessary number of infected cases. As opposed to a transition from S to E_I, a transition from the compartment S to I can be triggered by symptom development. This is explained by the fact that the individuals in the compartment I are only partly symptomatic.

No isolation rate is necessary because it is presupposed that the confirmed infected cases i.e. all people in the compartment I are isolated.

Since the predicted number of individuals in hospital or intensive care is of strong interest for planning bed and personnel capacities and data are available here, H

and C are treated as a single compartment each in the implementation realized in this thesis.

5.2   Systems of Ordinary Differential Equations

The mathematical formulation of systems of ODEs is briefly introduced in Subsection 5.2.1 for the purpose of embedding the compartment model theory of Chapters 3 and 4 into an initial value problem, in which the state variables representing the compartment sizes depend on certain partly non-constant parameters. State variables are used to describe dynamic systems. In the Subsections 5.2.2. to 5.2.3, the dynamical models and corresponding systems of ODEs of the $SVIHCDR$ model and $SV\tilde{A}IHCDR$ model are presented. A model compose of three age groups is explained in Subsection 5.2.4.

In the $MATLAB$ implementations of Chapter 6, a $SIHCDR$, which is the $SVIHCDR$ model without a compartment V, is applied to data of the countries Germany and Sweden. An all-or-nothing-vaccinated compartment is then implemented within the $SVID$ age group model based on German data.

5.2.1   Formulation of the Initial Value Problem

In the sequel, the term ODE model is referred to as the development of states variables of an s-dimensional system over time. Let an epidemic dynamic ODE model, e.g. the SARS-CoV-2-fitted model, be given. The target is the integration of the correspondent system of differential equations from a point in time t_0 to t_l with initial conditions x_0, where x_0 refers to the initial compartment sizes, in order to obtain a time series of compartment size data. Then the following methodological and mathematical precautions should be taken.

A system of ODEs is usually stated in the form of an initial value problem as in Equation (5.1).

$$x'(t) = F(t, x, \vartheta), \quad x(t_0) = x_0, \quad x(t) \in \mathbb{R}^s, \quad \vartheta \in \mathbb{R}^m, \quad t \in [t_0, t_l], \quad (5.1)$$

where $x_i(t)$ symbolizes the size of the compartment $\mathcal{K}_i$ at time instant $t, i \in \{1, ..., s\}$, in the case of a compartment model, and the initial conditions referring to an initial point in time t_0 are given by

$$x_0 = [x_0^{(1)}, x_0^{(2)}, ..., x_0^{(s)}]^\top.$$

The vector of all parameters of partly adjustable size occurring in the system of ODEs is given by

$$\vartheta = [\vartheta_1, ..., \vartheta_m]^\top \in \mathbb{R}^m.$$

In the case of a compartment model the equations $x_i'(t) = F(t, x_i, \vartheta)$, $i \in \{1, ..., s\}$, are nonlinear, i.e. Equation (5.1) represents a nonlinear system. It is assumed that the function $F : \mathbb{R} \times \mathbb{R}^s \times \mathbb{R}^m \to \mathbb{R}^s$ is sufficiently smooth i.e. normally C^3. Piecewise Lipschitz continuity is sufficient to guarantee the existence of a solution [245]. Supposing a unique solution vector exists for (5.1), numerical approximation schemes can be used in order to obtain a solution trajectory $x^*(t; \vartheta)$.

The $MATLAB$ solver $ode45$, that is applied in Chapter 6 to solve (5.1), was introduced in the late 1990s and is based on an algorithm of Dormand and Prince. It is a Runge–Kutta method and uses 6 stages, provides fourth and fifth order formulas, has local extrapolation and a companion interpolant [246]. Like the $MATLAB$ solver $ode23$, $ode45$ is designed for nonstiff systems of differential equations, where $ode23$ can be more efficient at problems with crude tolerances or present moderate stiffness. As opposed to this, the solver $ode15s$ should be used if $ode45$ is inefficient and it is suspected the underlying problem is stiff. The solver $ode23t$ can solve differential algebraic equations (DAEs) and is the first choice if the underlying problem is moderately stiff and a solution without numerical damping is wanted. The solver $ode15i$ is designed for fully implicit problems [247].

5.2.2   The $SVIHCDR$ Model and the $SV\tilde{A}_I I_I HCDR$ Model

A model with a single infected compartment I can be established as indicated in Subsection 5.1.4. All individuals in the compartment I can be infectious and symptomatic, but can be in a state before or after symptom development or infectiousness. The rates of transition to and from this infected class have to be stated. The compartment S_q and compartments containing unconfirmed cases are omitted in the following for the reasons given in Subsections 5.1.1 and 5.1.3. The corresponding rate of transmission defined on the basis of the transmission rates of Subsection 4.3.3 is independent of an isolation rate and given in Equation (5.2).

$$\theta_I(t) = \beta \cdot \gamma(t) \cdot \left(1 - q(t)\right), \tag{5.2}$$

Thus the rate of transition from the compartment S to I is defined by

$$\Theta_I(t) := \theta_I(t) \cdot \left(\varepsilon_I \cdot I(t) + \varepsilon_H \cdot H(t) + \varepsilon_C \cdot C(t) \right),$$

where ε_I is the factor expressing the modification of the risk of transmitting the infection for individuals in the class I compared with those in the class I_U of the SARS-CoV-2-fitted model. Since the compartment I consists of confirmed symptomatic and confirmed pre-symptomatic cases and $\varepsilon_{I_I} > \varepsilon_{A_I}$ it holds that

$$\varepsilon_{I_U} > \varepsilon_{I_I} > \varepsilon_I .$$

If not all individuals in the compartment I are assumed to be infectious as opposed to the Basic SIR model, the factor ε_I must be selected as even smaller in this model. A time delay as explained in Subsection 4.3.4 is not incorporated in the model, since all individuals with a confirmed infection, i.e. exposed, asymptomatic infectious, symptomatic infectious, are comprised in the compartment I. The assumption that some of the cases pooled in the class I are only still asymptomatic and/or latently infected when they are reported.

Let the average time that individuals in the compartment I take from the confirmation of their contracted infection to recovery be $T_I \geq T$ units of time. It has to be as large as the period of contagiousness of not-hospitalized people if a direct detection and reporting of infected cases is assumed. It must be smaller than this period otherwise. The time T_I can be assumed to be not much larger than the time from symptom development to recovery, since most infections are not detected before symptoms become distinct. The individuals in the class I can be hospitalized owing to major symptoms instead of progressing to the compartment R. Thus the recovery rate is

$$w_2^I = \frac{1 - \mu \cdot T_I}{T_I} \cdot \left(1 - K \right)$$

and the hospitalization rate is

$$\eta^I = \frac{1 - \mu \cdot T_I}{T_I} \cdot K .$$

In the model, people in the class I cannot die due to their infection, but hospitalized ones can. The compartment S_q is excluded from the model in order to facilitate the later data search as proposed in Subsection 5.1.3.

A system of ODEs for the $SIVHCDR$ model can be established by regarding the inflow and outflow of each compartment. It emerges from the explained characteristic features of the seven compartments S, I, V, H, C, D and R as well as transition rates and is depicted in Table 5.1. In the systems of ODEs created in this section, a standard incidence is used. Thus in the multiplication of the transmission rate with the size of the susceptible compartment at time instant t, $S(t)$ is normalized by the size of the total population minus the size of the deceased compartment at time t. In the case of a $SVIHCDR$ model, it holds that $N(t) := S(t) + V(t) + I(t) + H(t) + C(t) + D(t) + R(t)$.

Table 5.1 System of Ordinary Differential Equations for the $SIVHCDR$ model displaying the population dynamics of 2019-nCoV

$$\frac{dS(t)}{dt} = \mathcal{V}_2 \cdot V(t) - \Theta_I(t) \cdot \frac{S(t)}{N - D(t)} - (\mu + \mathcal{V}_1) \cdot S(t) \ ,$$

$$\frac{dV(t)}{dt} = \mathcal{V}_1 \cdot S(t) - (\mu + \mathcal{V}_2) \cdot V(t) \ ,$$

$$\frac{dI(t)}{dt} = \Theta_I(t) \cdot \frac{S(t)}{N - D(t)} - \left(\omega_2^I + \eta^I + \mu\right) \cdot I(t) \ ,$$

$$\frac{dH(t)}{dt} = \eta^I \cdot I(t) - \left(\omega_3 + \lambda_1 + \xi + \mu\right) \cdot H(t) \ ,$$

$$\frac{dC(t)}{dt} = \xi \cdot H(t) - \left(\omega_4 + \lambda_2 + \mu\right) \cdot C(t) \ ,$$

$$\frac{dD(t)}{dt} = \lambda_1 \cdot H(t) + \lambda_2 \cdot C(t) \ ,$$

$$\frac{dR(t)}{dt} = \omega_2^I \cdot I(t) + \omega_3 \cdot H(t) + \omega_4 \cdot C(t) - \mu \cdot R(t) \ .$$

The dynamics of the model with pooled infected compartments emerging from Table 5.1 can be visualized by a transition and transmission diagram, which is presented in Figure 5.1. Blue arrows from one to another compartment indicate a transition, whereas the compartment, from which a red dashed arrow originates, can infect susceptibles.

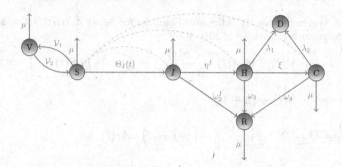

Figure 5.1 Compartment model for SARS-CoV-2 with one infected compartment, the non-observance of a quarantine compartment and an all-or-nothing vaccine

It was described in Subsection 5.1.4 that the classes E_I and A_I can be condensed to an asymptomatic compartment $\tilde{A}_I$ instead of pooling the classes E_I, A_I and I_I to a single compartment I. Proceeding from the $SVIHCDR$ model now, the class I can be divided into the asymptomatic compartment $\tilde{A}_I$ and the symptomatic class I_I. The aim of this step is the ability to predict the future numbers of asymptomatic individuals separated from symptomatic cases. The rates ω_2 and η are used for modelling instead of ω_2^I and η^I. All recovery, hospitalization and disease-induced CFRs depend on T_I instead of T.

The rate of transition from the compartment S to $\tilde{A}_I$ is defined as

$$\Theta_{\tilde{A}_I}(t) := \theta_I(t) \cdot \left(\varepsilon_{\tilde{A}_I} \cdot \tilde{A}_I + \varepsilon_{I_I} \cdot I_I(t) + \varepsilon_H \cdot H(t) + \varepsilon_C \cdot C(t) \right),$$

where $\varepsilon_{\tilde{A}_I}$ is the factor expressing the modification of the risk of transmitting the infection for individuals in the class $\tilde{A}_I$ compared to those in the class I_U of the SARS-CoV-2-fitted model.

In the case of a $SV\tilde{A}_I IHCDR$ model, it holds that $N(t) := S(t) + V(t) + \tilde{A}_I(t) + I(t) + H(t) + C(t) + D(t) + R(t)$.

The system of ODEs of the $SV\tilde{A}_I IHCRD$ model is presented in Table 5.2.

Table 5.2 System of Ordinary Differential Equations for the $SV\tilde{A}_I I_I HCDR$ model displaying the population dynamics of 2019-nCoV

$$\frac{dS(t)}{dt} = \mathcal{V}_2 \cdot V(t) - \Theta_{\tilde{A}_I}(t) \cdot \frac{S(t)}{N - D(t)} - (\mu + \mathcal{V}_1) \cdot S(t)\ ,$$

$$\frac{dV(t)}{dt} = \mathcal{V}_1 \cdot S(t) - (\mu + \mathcal{V}_2) \cdot V(t)\ ,$$

$$\frac{d\tilde{A}_I(t)}{dt} = \Theta_{\tilde{A}_I}(t) \cdot \frac{S(t)}{N - D(t)} - \left(\chi_{II} + \mu\right) \cdot \tilde{A}_I(t)\ ,$$

$$\frac{dI_I(t)}{dt} = \chi_{II} \cdot \tilde{A}_I(t) - \left(\omega_2 + \eta + \dot{\mu}\right) \cdot I_I(t)\ ,$$

$$\frac{dH(t)}{dt} = \eta \cdot I_I(t) - \left(\omega_3 + \lambda_1 + \xi + \mu\right) \cdot H(t)\ ,$$

$$\frac{dC(t)}{dt} = \xi \cdot H(t) - \left(\omega_4 + \lambda_2 + \mu\right) \cdot C(t)\ ,$$

$$\frac{dD(t)}{dt} = \lambda_1 \cdot H(t) + \lambda_2 \cdot C(t)\ ,$$

$$\frac{dR(t)}{dt} = \omega_2 \cdot I_I(t) + \omega_3 \cdot H(t) + \omega_4 \cdot C(t) - \mu \cdot R(t)\ .$$

5.2.3 The SARS-CoV-2-fitted Model

The SARS-CoV-2-fitted model including all 13 equations presented in Table 5.3. In the case of this enhanced model, it holds that

$$N(t) := S(t) + V(t) + S_q(t) + E_I(t) + E(t) + A_I(t) + A(t) + I_I(t) + I_U(t)$$
$$+ H(t) + C(t) + D(t) + R(t).$$

Table 5.3 System of Ordinary Differential Equations for the SARS-CoV-2-fitted model displaying the population dynamics of 2019-nCoV

$$\frac{dS(t)}{dt} = \mathcal{V}_2 \cdot V(t) - \left(\Theta_{S_q}(t) + \Theta_{E_I}(t) + \Theta_E(t) \right) \cdot \frac{S(t)}{N - D(t)} + \hat{\phi} \cdot S_q(t) - \mu \cdot S(t) \ ,$$

$$\frac{dV(t)}{dt} = \mathcal{V}_1 \cdot S(t) - (\mu + \mathcal{V}_2) \cdot V(t) \ ,$$

$$\frac{dS_q(t)}{dt} = \Theta_{S_q}(t) \cdot \frac{S(t)}{N - D(t)} - \left(\hat{\phi} + \mu \right) \cdot S_q(t) \ ,$$

$$\frac{dE_I(t)}{dt} = \Theta_{E_I}(t) \cdot \frac{S(t)}{N - D(t)} - (\psi_{II} + \mu) \cdot E_I(t) \ ,$$

$$\frac{dE(t)}{dt} = \Theta_E(t) \cdot \frac{S(t)}{N - D(t)} - (\psi_U + \psi_I + \mu) \cdot E(t) \ ,$$

$$\frac{dA_I(t)}{dt} = \psi_{II} \cdot E_I(t) + \psi_I \cdot E(t) - \left(\chi_{II} + \mu \right) \cdot A_I(t) \ ,$$

$$\frac{dA(t)}{dt} = \psi_U \cdot E(t) - \left(\chi_U + \chi_I + \mu \right) \cdot A(t) \ ,$$

$$\frac{dI_I(t)}{dt} = \chi_{II} \cdot A_I(t) + \chi_I \cdot A(t) + \zeta \cdot I_U(t) - \left(\omega_2 + \eta + \mu \right) \cdot I_I(t) \ ,$$

$$\frac{dI_U(t)}{dt} = \chi_U \cdot A(t) - \left(\omega_1 + \zeta + \lambda_3 + \mu \right) \cdot I_U(t) \ ,$$

$$\frac{dH(t)}{dt} = \eta \cdot I_I(t) - \left(\omega_3 + \xi + \lambda_1 + \mu \right) \cdot H(t) \ ,$$

$$\frac{dC(t)}{dt} = \xi \cdot H(t) - \left(\omega_4 + \lambda_2 + \mu \right) \cdot C(t) \ ,$$

$$\frac{dD(t)}{dt} = \lambda_1 \cdot H(t) + \lambda_2 \cdot C(t) \ ,$$

$$\frac{dR(t)}{dt} = \omega_1 \cdot I_U(t) + \omega_2 \cdot I_I(t) + \omega_3 \cdot H(t) + \omega_4 \cdot C(t) - \mu \cdot R(t) \ .$$

The dynamics of the enhanced model expressed by Table 5.3, can be visualized by a transition and transmission diagram, which is presented in Figure 5.2.

Once again, the compartment, from which a red dashed arrow originates, can infect susceptibles.

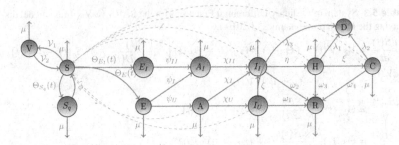

Figure 5.2 The SARS-CoV-2-fitted compartment model with an all-or-nothing vaccine

The above model realistically depicts the population dynamics triggered by the SARS-CoV-2 pandemic.

It includes the control strategies of quarantine and isolation as well as the 5 distinct stages of disease progression of latent infection, asymptomatic infectiousness, symptomatic contagiousness, hospitalization and intensive care, where the first three states are separated into an isolated (confirmed) and an unisolated (unconfirmed) compartment. Nevertheless, this detailed model is not implemented in this thesis for reasons related to data availability, that were explained in Section 5.1.

5.2.4   Age Group Model

Interactions among individuals of the same or different age groups can be included into a compartment model. Known characteristics of the COVID-19 outbreak prefigure the heterogeneous distribution of cases with respect to the age structure of the population. Subsequently, age dependencies within subpopulations are taken into consideration here instead of geographic inhomogeneities [78]. Distinguishing between people in different age groups allows the model to better characterize contacts between individuals and to fine-tune the effect of intervention measures on contacts reduction [78]. For example, child-child contacts differ from senior-adults in their frequency as wells as intensity.

According to the analysis of Figure 2.3 a reasonable realization of age groups in a model displaying the SARS-CoV-2 population dynamics incorporates the distinction 3 or more age groups. The division of each compartment into a class of children of $0 - 19$ years (group $i = 1$), adults of $20 - 59$ years (group $i = 2$) and people of 60 years or older (group $i = 3$) seems justifiable as the purpose of age group inclusion is the following:

The ability to distinguish between the social behaviour and other factors influencing the transmission risk emerging from a certain age group as well as the vulnerability, contagiosity and risk of showing major symptoms or being hospitalized of the age group.

The $SVIHCDR$ model is taken as a basis of age group model development here. This also means that the accessibility of an all-or-nothing vaccine but not a leaky vaccine is assumed. It turns into the $S_j V_j I_j H_j C_j D_j R_j$ model under the inclusion of age groups $j \in \{1, 2, 3\}$. Consequently, there are three individual compartment models, that are coupled by contact rates and disease transmission among individuals of different age groups [78]. It is defined that a compartment of the $SVIHCDR$ model has the same size as the sum of the three age group compartments with the same characteristics in the $S_j V_j I_j H_j C_j D_j R_j$ model:

$$\mathcal{K}(t) = \sum_{j=1}^{3} \mathcal{K}_j(t) \quad \forall \mathcal{K}_j \in \{S, V, I, H, C, D, R\}.$$

Let $\forall t \ \hat{X}_j(t) = \varepsilon_{I_j} \cdot I_j(t) + \varepsilon_{H_j} \cdot H_j(t) + \varepsilon_{C_j} \cdot C_j(t)$ in the case of the given $S_j V_j I_j H_j C_j D_j R_j$ model as well as the rate $\gamma_{ji}(t)$ be a function representing the average number of contacts between an individual of the infected age group I_j with one of the susceptible group S_i per unit of time. It follows for the sum of modification factors multiplied by compartment sizes (that is $\hat{X}(t)$) and the contact rate of a compartment $\mathcal{K}_i$ (that is $\gamma_i(t)$) that

$$\hat{X}(t) = \sum_{j=1}^{3} \hat{X}_j(t) \quad \text{and} \quad \gamma_i(t) = \sum_{j=1}^{3} \gamma_{ji}(t).$$

It holds that $\gamma_{ji}(t) = \gamma_{ij}(t)$ for all $i, j \in \{1, 2, 3\}$, such that the matrix $\gamma \in \mathbb{R}^{3 \times 3}$ is symmetric. Let the function $q_i(t)$ describe the ratio of individuals of the age group i who are put in quarantine per unit of time t, and β_{ji} represent the general risk of a transmission emerging from the age group j infecting the age group i.

A transmission rate $\theta_{Iji}(t), i \in \{1, 2, 3\}$ concerns the transmission of the infection from an age group j to i.

It holds that

$$\theta_{Iji}(t) = \beta_{ji} \cdot \gamma_{ji}(t) \cdot \left(1 - q_i(t)\right),$$

where the case $j = i$ is possible as a person can infect another person of the same age group. Subsequently, the rate of transition from a compartment S_i to I_i is

$$\Theta_{I_i}(t) = \left(\sum_{j=1}^{3} \theta_{I_{ji}}(t)\right) \cdot \hat{X}(t) = \left(\sum_{j=1}^{3} \beta_{ji} \cdot \gamma_{ji}(t)\right) \cdot \left(1 - q_i(t)\right) \cdot \hat{X}(t).$$

Regarding the adoption of the transition rates μ, ξ, $\omega_2(t)$, $\eta_2(t)$, $\omega_3(t)$, $\omega_4(t)$, $\lambda_1(t)$ and $\lambda_2(t)$ to the different age groups, the system of ODEs for the corresponding age group compartment model is established in Table 5.4.

Table 5.4 System of Ordinary Differential Equations for a model displaying the population dynamics of 2019-nCoV including three age groups expressed by the indices $i, j \geq 1, i, j \in \{1, 2, 3\}$ ($S_i V_i I_i H_i C_i D_i R_i$ model)

$$\frac{dS_i(t)}{dt} = \mathcal{V}_{2i} \cdot S_i(t) - \Theta_{Ii}(t) \cdot \frac{S_i(t)}{N - D(t)} - (\mu_i + \mathcal{V}_{1i}) \cdot S_i(t) \quad,$$

$$\frac{dV_i(t)}{dt} = \mathcal{V}_{1i} \cdot S_i(t) - (\mu_i + \mathcal{V}_{2i}) \cdot V_i(t) \quad,$$

$$\frac{dI_i(t)}{dt} = \Theta_{Ii}(t) \cdot \frac{S_i(t)}{N - D(t)} - \left(\omega_{2i} + \eta_i + \mu_i\right) \cdot I_i(t) \quad,$$

$$\frac{dH_i(t)}{dt} = \eta_i \cdot I_i(t) - \left(\omega_{3i} + \lambda_{1i} + \mu_i\right) \cdot H_i(t) \quad,$$

$$\frac{dC_i(t)}{dt} = \xi_i \cdot H_i(t) - \left(\omega_{4i} + \lambda_{2i} + \mu_i\right) \cdot C_i(t) \quad,$$

$$\frac{dD_i(t)}{dt} = \lambda_{1i} \cdot H_i(t) + \lambda_{2i} \cdot C_i(t) \quad,$$

$$\frac{dR_i(t)}{dt} = \omega_{2i} \cdot I_i(t) + \omega_{3i} \cdot H_i(t) + \omega_{4i} \cdot C_i(t) - \mu_i \cdot R_i(t) \quad.$$

Owing to the age ranges of the three selected age groups the following relations between their average parameter sizes seem reasonable:

$\mu_3 > \mu_2 > \mu_1$,

$\omega_{k3} < \omega_{k2} \leq \omega_{k1}$ for $k = 2, 3, 4$,

$\eta_3 > \eta_2 \geq \eta_1$,

$\xi_3 > \xi_2 \geq \xi_1$,

$\lambda_{s3} > \lambda_{s2} > \lambda_{s1}$ for $s = 1, 2$,

and due to an assumed vaccination policy of firstly vaccinating the elder generation (> 60 years)

$\mathcal{V}_{13} > \mathcal{V}_{12} \geq \mathcal{V}_{11}$

and due to an assumed stronger waning protective effect of vaccinations in the elder generation

$$v_{23} > v_{22} \geq v_{21} \, .$$

For the contact rates, which influence the transmission rates, logical considerations with respect to the numbers of contacts between the different age groups make it reasonable that

$$\gamma_{13} < \gamma_{11} < \gamma_{12}, \gamma_{21} < \gamma_{23} < \gamma_{22}, \gamma_{31} < \gamma_{33} < \gamma_{32} \, , \text{ and more exactly}$$
$$\gamma_{13} < \gamma_{11} < \gamma_{33} < \gamma_{12} < \gamma_{23} < \gamma_{22}.$$

The dynamics between a $S_i I_i H_i C_i D_i R_i$ model and a $S_j V_j I_j H_j C_j D_j R_j$ model, where $j \neq i$, can be visualized by a transition and transmission diagram, which is presented in Figure 5.3. A blue arrow from one to another compartment implies a transition within an age group, whereas the compartment, from which a red dashed arrow originates, can infect susceptibles in the same age group. A purple dashed arrow indicates a transmission from the age group i to j, whereas a brown dashed arrow indicates a transmission from the age group j to i.

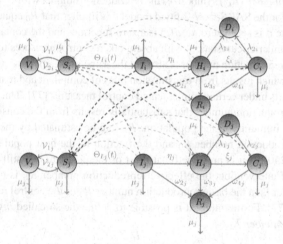

Figure 5.3 Transition dynamics between two age groups in an age-structured compartment model for SARS-CoV-2 with age group-coupled transmission dynamics.

An age group model including less compartments than the $S_i V_i I_i H_i C_i D_i R_i$ model is implemented in $MATLAB$. The process of implementation is explained and results are discussed in Chapter 6.

5.3   Reproduction Numbers

Reproduction numbers are indices of infectious diseases that are used as measures of transmissibility. There are definitions of different kinds of reproduction numbers, of which 3 are introduced in this thesis.

The *basic reproduction number* (R_0) is the reproduction number that is mentioned most often with regard to infectious diseases and can be described as the most important quantity in infectious disease epidemiology [195, p. 873]. It is defined as the expected number of secondary infections effected by the first infected individual introduced into a population of solely susceptible individuals [250]. Furthermore, it is an indicator of the infective potential of a pathogen and is the crucial index for the pandemic potential. The zero in R_0 stands for the fact that no person in the model is assumed to be immune against the regarded infection. As we will see in the sequel, R_0 depends on the considered model to describe the pandemic.

R_0 is defined in the absence of any control measures. In contrast to this, the *control reproduction number* (R_C) symbolizes the reproduction number with control actions taken [251]. For the SARS-CoV-2-fitted model this implies that R_0 equals R_C if the quarantine rate q is set to zero, or $q(t)$ is a zero function, and the contact rate $\gamma(t)$ is as high as under circumstances with absent taken control measures for all t.

The basic reproduction number is also not the same as the *effective reproduction number* R_E, which results from transmission rates of compliant and non-compliant population parts under certain implemented control measures [77]. This means that R_E is the reproduction number that substantially results from the consideration of an increasing immunity in the population. R_E can be estimated by the product of the basic reproduction number R_0 and the fraction of the host population that is susceptible [250]. If it is presumed that most of the population is still susceptible at the time of intervention, the effective reproduction number R_E is usually only slightly smaller than the basic reproduction number R_0 or the control reproduction number R_C [77]. Furthermore, it is possible to define the so-called *instantaneous reproduction number* R_t:

$$R_t = \frac{I_t}{\sum_{k=1}^{t} I_{t-k} \cdot w_k},$$

where w_k is the discrete probability distribution of the serial interval concerning secondary infections effected by the individual k, $k \in \{1, 2, 3, ...\}$ [252]. Thus it is assumed that the cases I_t newly infected at time t contracted the infection from a fraction w_k of those infected at time $t - k$.

The number R_t can be regarded over a time interval $\mathcal{T}$ instead of a point in time t, where additionally the moving average of the number of new infections over τ_n days is incorporated and equals the numerator of the fraction below. The instantaneous reproduction number over the interval $\mathcal{T}$ is defined as [252]

$$R_{t,\mathcal{T}} = \frac{\frac{1}{\tau_n}\sum_{s=t-\mathcal{T}+1}^{t} I_s}{\frac{1}{\tau_n}\sum_{s=t-\mathcal{T}+1}^{t}\sum_{k=1}^{s} I_{t-k}w_k}.$$

The German RKI uses this definition in order to compute the reproduction number of 2019-nCoV. It utilizes the assumption that

$$\sum_{s=t-\mathcal{T}+1}^{t}\sum_{k=1}^{s} I_{t-k}\cdot w_k = \sum_{s=t-\mathcal{T}+1}^{t} I_{s-\mathcal{T}^I} \tag{5.3}$$

with $\mathcal{T}^I$ the average length of the incubation period [252]. The reproduction number of SARS-CoV-2 in Germany reached a maximum of 2.88 on June 21st 2020 as well as a smaller local maximum of 1.52 on January 11th 2021 with $\mathcal{T}$ set to the value 4 [253].

For an arbitrary system of ODEs of a compartment model the basic and control reproduction numbers can be computed with the aid of the technique of so-called *next generation matrices* (NGMs) [195].

Using the function $f : \mathbb{R}^n \to \mathbb{R}^n$, that maps the state variables to their derivations, the dynamics of the system of ODEs can be written as

$$X'(t) = f\big(X(t)\big), \quad \text{with } X = (X_1, ..., X_p, X_{p+1}, ..., X_n)^{\top}$$
$$\text{the vector of compartment variables,}$$

of which $X_{p+1}, ..., X_n$ are the infected states. The first step to obtain R_0 is to linearise the infected subsystem about the infection-free steady state that exists as a rule [195, p. 874]. In an epidemiological sense, the linearisation reflects that R_0 characterizes the potential for an initial spread of an infectious agent when it is introduced into a fully susceptible population [195, p. 874]. Therefore, the ODEs of the system are separated into the state variables and the entering fluxes related to the infectious process. Let $\mathcal{F}_i$ be the flux of newly infected individuals in the compartment i, and $\mathcal{V}_i^+$ ($\mathcal{V}_i^-$) the other entering (leaving) fluxes related to the compartment i, $i \in \{1, ..., n\}$. All three of them are non-negative functions.

Then the infected subsystem can be separated as [254]

$$f(X) = \mathcal{F}(X) + \mathcal{V}^+(X) + \mathcal{V}^-(X).$$

Hence the system is decomposed into the matrices $\mathcal{F}$ and $\mathcal{V} = \mathcal{V}^+ + \mathcal{V}^-$ at first. An endemic equilibrium (EE) point is a steady-state solution where the disease persists in the population, which is the case when all state variables X_i are positive and $R_0 > 1$. As opposed to this, a disease-free equilibrium (DFE) point of a system of ODEs corresponding to a compartment model is a steady-state solution where there is no disease. It exists in the absence of the spread of the infection, which is for $R_0 < 1$. A DFE point is given by $X^* = (X_1^*, ..., X_p^*, 0..., 0)$, where the zero appears $n - p$ times, for which it holds that [254]

$$D_{x^*}(\mathcal{F}) = \begin{pmatrix} 0 & 0 \\ 0 & \mathcal{E} \end{pmatrix}, \quad D_{x^*}(\mathcal{V}) = \begin{pmatrix} J_1 & J_2 \\ 0 & T \end{pmatrix}.$$

with J_1 and J_2 2 resulting matrices. It holds that

$$f(X_i) = X_i' = \mathcal{F}_i(X) + \mathcal{V}_i(X) \quad \forall i \in \{p+1, ..., n\}$$

and the linearised system at the DFE can be written by means of the linearisation of $\mathcal{F}$ and $\mathcal{V}$:

$$\mathcal{E}_{ij} = \frac{\delta \mathcal{F}_i}{\delta x_j}(X^*), \quad T_{ij} = -\frac{\delta \mathcal{V}_i}{\delta x_j}(X^*).$$

Consequently, it holds that $X' = (\mathcal{E} + T) \cdot X$ [255]. The matrix T corresponds to the transmissions and the matrix $\mathcal{E}$ to the transitions in the system. An entry T_{ij} is the rate at which individuals in the infected state j give rise to individuals in the infected state i, and an entry $\mathcal{E}_{ij}$ is the rate at which individuals transit from the compartment j to i, that is not initiated by transmission [195, p. 875]. Subsequently, all epidemiological events effecting new infections are included via T, whereas all other transition events are incorporated via $\mathcal{E}$. Progress to either death or immunity guarantees that $\mathcal{E}$ is invertible [195].

Let the spectrum of any square matrix Q be denoted by $\sigma(Q)$, the *spectral radius* of Q be defined by

$$\rho(Q) = \max\left\{|\lambda|, \lambda \in \sigma(Q)\right\},$$

and the stability modulus of Q by

$$\alpha(Q) = \max\Big\{Re(\lambda), \lambda \in \sigma(Q)\Big\}.$$

If $\alpha(T) < 0$, then the basic reproduction number R_0 linked to the DFE X^* of the underlying system of ODEs is defined as [254]

$$R_0 = \rho\Big(\mathcal{E} \cdot (-T^{-1})\Big) = \rho\Big(-\mathcal{E} \cdot T^{-1}\Big), \quad \text{where } K_L := -\mathcal{E} \cdot T^{-1}.$$

The matrix K_L is called the NGM with large domain.

The *Poincaré-Lyapunov linearisation theorem* states [254]:

Let X^* be a steady state of $X'(t) = f(X(t))$, where $f : \mathbb{R}^n \to \mathbb{R}^n$ is locally Lipschitz-continuous. Then X^* is asymptotically stable if $\alpha\Big(D_{x^*}(f)\Big) < 0$ and unstable if $\alpha\Big(D_{x^*}(f)\Big) > 0$.

Following the partition $f = \mathcal{F} + \mathcal{V}$, it has to be focused on the sign of $\alpha(\mathcal{E} + T)$ to prove the stability of the DFE. If $R_0 > 1$, it holds that $\alpha(\mathcal{E} + T) > 0$ such that $\alpha\Big(D_{x^*}(\mathcal{F} + \mathcal{V})\Big) > 0$, and the respective DFE X^* is unstable. Analogously, it is asymptotically stable if $R_0 < 1$ under the condition $\alpha(J_1) < 0$ [254]. In the case of the existence of multiple disease-free equilibria, it can be possible to calculate a R_0-value related to each equilibrium, but the definition of a global R_0 is much more complicated and requires a case-by-case study [254].

Equivalent to K_L, the so-called next-generation matrix (NGM) with classical domain can be defined as

$$K_C = E^\top \cdot \mathcal{E} \cdot T^{-1} \cdot E,$$

where E is the unity matrix of the corresponding dimension. It can be proven that the NGM with classical domain and the NGM with large domain have the same non-zero eigenvalue [256, p. 261]:

Let $\hat{v}$ be an eigenvector of K_C with corresponding eigenvalue λ. Then

$$K_C \cdot \hat{v} = E^\top \cdot \mathcal{E} \cdot T^{-1} \cdot E \cdot \hat{v} \overset{!}{=} \lambda \cdot \hat{v}.$$

Multiplying this identity by E yields

$$E \cdot E^\top \cdot \mathcal{E} \cdot T^{-1} \cdot E \cdot \hat{v} = E \cdot \lambda \cdot \hat{v},$$

where

$$E \cdot E^\top \cdot \mathcal{E} = \mathcal{E} .$$

Thus $E \cdot \hat{v}$ is an eigenvector of K_L, too, with the corresponding eigenvalue λ.

The matrix K_C has a lower dimension than K, making the computation of R_0 from K_C easier and increasing the possibility of obtaining an explicit expression [195, p. 874]. The NGM with large domain typically uses the dynamics of more states to describe the evolution of infection generations than K_C [195, p. 874]. Solely the infected states are involved in the action of K_C. An entry K_{Cij} is the expected number of new cases with state-of-infection i generated by an individual who has just entered the state-of-infection j [195, p. 874].

The basic reproduction number R_0 for the $SVIHCDR$ model is computed in this thesis. The infected states as well as states-of-infectiousness were defined in Section 4.3. In the $SVIHCDR$ model, the compartments I, H and C are the infected states, which are also regarded as infectious in this section. The correspondent system of ODEs can be found in Table 5.1.

Table 5.1 shows that the variables D and R do not appear in any other equation than the differential equation describing their own respective change over time. The variable V exclusively occurs in the first and second equation in Table 5.1 describing the change of the susceptible or vaccinated compartment over time, respectively. Moreover, the variable S explicitly appears in only the second equation in Table 5.1, that describes the change of the vaccinated compartment over time. It implicitly occurs in the third equation in Table 5.1, that describes the change of the infected compartment over time, via the term $\Theta_I(t) \cdot \frac{S(t)}{N-D(t)}$ in the case of a saturated incidence, which becomes $\Theta_I(t)$ if S is set to $N - D(t)$ [195, p. 875].

For these reasons, the ODEs for S, V, D and R can be omitted in the computations of the basic reproduction number R_0. The system of ODEs consisting of exclusively the ODEs for I, H and C can be referred to as the *linearised infected subsystem* of the actual system of ODEs because they only describe the production of new infections and changes in the states of already existing infected cases [195, p. 875]. The NGM that emerges from the separation of the transition rates of this infected subsystem into the matrices $\mathcal{E}$ and T is of classical domain. In the sequel, it is computed like the NGM with large domain K_L, since uninfected states are omitted already in the creation of the matrices $\mathcal{E}$ and T.

In the following computations, the time reference t is dropped for simplicity, the first row or column refers to the compartment I, the second one to H and the third one to C, respectively. The transmission matrix for the infected subsystem of the $SVIHCDR$ model is given by

$$\mathcal{E} = \begin{pmatrix} \beta \cdot \gamma \cdot (1-q) \cdot \varepsilon_I & \beta \cdot \gamma \cdot (1-q) \cdot \varepsilon_H & \beta \cdot \gamma \cdot (1-q) \cdot \varepsilon_C \\ 0 & 0 & 0 \\ 0 & 0 & 0 \end{pmatrix}$$

the respective transition matrix by

$$T = \begin{pmatrix} -(\mu + \omega_2 + \eta) & 0 & 0 \\ \eta & -(\mu + \xi + \omega_3 + \lambda_1) & 0 \\ 0 & \xi & -(\mu + \lambda_2 + \omega_4) \end{pmatrix}.$$

The eigenvalues of the next generation matrix $K_L = -T \cdot \mathcal{E}^{-1}$ are obtained via $MATLAB$. The first two eigenvalues are zero. The basic reproduction number R_0 equals the third and largest eigenvalue of K_L, which is defined as the spectral radius of K_L:

$$\begin{aligned} \rho(K_L) &= \rho(-T \cdot \mathcal{E}^{-1}) \\ &= -\left(\frac{\varepsilon_I \cdot \theta_I}{\eta + \mu + \omega_2} + \frac{\varepsilon_H \cdot \eta \cdot \theta_I}{(\eta + \mu + \omega_2) \cdot (\lambda_1 + \mu + \omega_3 + \xi)} \right. \\ &\quad \left. + \frac{\varepsilon_C \cdot \eta \cdot \theta_I \xi}{(\eta + \mu + \omega_2) \cdot (\lambda_2 + \mu + \omega_4) \cdot (\lambda_1 + \mu + \omega_3 + \xi)} \right). \end{aligned}$$

The normalized forward sensitivity index is used for a sensitivity analysis of the basic reproduction number R_0 depending on a certain model parameter. This index is defined as [257]

$$\mathcal{S}_p^{R_0} = \frac{\partial R_0}{\partial p} \cdot \frac{p}{R_0}, \tag{5.4}$$

where p is a selected model parameter. A positive (negative) sensitivity index means that the prevalence of the disease increases (decreases) if the value of the respective parameter is increased. For instance, a sensitivity index $\mathcal{S}_p^{R_0} = 0.5$ means that R_0 would increase (decrease) by 5% if the parameter p was increased (decreased) by 10%. In Table 5.5, the sensitivity index with reference to the model parameter $p \in \{\gamma, q, \beta, \eta, \xi\}$ is depicted.

Table 5.5 Sensitivity analysis of model parameters

$\dot{\gamma}$ $\quad \frac{\delta R_0}{d\gamma} \cdot \frac{\gamma}{R_0} = -\frac{\beta \varepsilon_I (q-1)}{(\eta+\mu+\omega_2)} - \frac{\beta \varepsilon_H \eta (q-1)}{(\eta+\mu+\omega_2)(\lambda_1+\mu+\omega_3+\xi)} - \frac{\beta \varepsilon_C \eta \xi (q-1)}{(\eta+\mu+\omega_2)(\lambda_2+\mu+\omega_4)(\lambda_1+\mu+\omega_3+\xi)}$

q $\quad \frac{\delta R_0}{dq} \cdot \frac{q}{R_0} = -\frac{\beta \gamma \varepsilon_I}{\eta+\mu+\omega_2} - \frac{\beta \gamma \varepsilon_H \eta}{(\eta+\mu+\omega_2)(\lambda_1+\mu+\omega_3+\xi)} - \frac{\beta \gamma \varepsilon_C \eta \xi}{(\eta+\mu+\omega_2)(\lambda_2+\mu+\omega_4)(\lambda_1+\mu+\omega_3+\xi)}$

β $\quad \frac{\delta R_0}{d\beta} \cdot \frac{\beta}{R_0} = -\frac{\gamma \varepsilon_I (q-1)}{(\eta+\mu+\omega_2)} - \frac{\gamma \varepsilon_H \eta (q-1)}{(\eta+\mu+\omega_2)(\lambda_1+\mu+\omega_3+\xi)} - \frac{\gamma \varepsilon_C \eta \xi (q-1)}{(\eta+\mu+\omega_2)(\lambda_2+\mu+\omega_4)(\lambda_1+\mu+\omega_3+\xi)}$

η $\quad \frac{\delta R_0}{d\eta} \cdot \frac{\eta}{R_0} = \frac{\beta \gamma \varepsilon_I (q-1)}{(\eta+\mu+\omega_2)^2} - \frac{\beta \gamma \varepsilon_H (q-1)}{(\eta+\mu+\omega_2)(\lambda_1+\mu+\omega_3+\xi)} + \frac{\beta \gamma \varepsilon_H \eta (q-1)}{(\eta+\mu+\omega_2)^2(\lambda_1+mu+\omega_3+\xi)}$

$\qquad\qquad - \frac{\beta \gamma \varepsilon_C (q-1)}{(\eta+\mu+\omega_2)(\lambda_2+\mu+\omega_4)(\lambda_1+\mu+\omega_3+\xi)} + \frac{\beta \gamma \varepsilon_C \eta (q-1)}{(\eta+\mu+\omega_2)^2(\lambda_2+\mu+\omega_4)(\lambda_1+\mu+\omega_3+\xi)}$

ξ $\quad \frac{\delta R_0}{d\xi} \cdot \frac{\xi}{R_0} = \frac{\beta \gamma \varepsilon_H \eta (q-1)}{(\eta+\mu+\omega_2)(\lambda_1+\mu+\omega_3+\xi)^2} - \frac{\beta \gamma \varepsilon_C \eta (q-1)}{(\eta+\mu+\omega_2)(\lambda_2+\mu+\omega_4)(\lambda_1+\mu+\omega_3+\xi)}$

$\qquad\qquad + \frac{\beta \gamma \varepsilon_C \eta \xi (q-1)}{(\eta+\mu+\omega_2)(\lambda_2+\mu+\omega_4)(\lambda_1+\mu+\omega_3+\xi)^2}$

Parameter values obtained from the implementation of the $SVIHCDR$ model and associated parameter optimization in Chapter 6 can be plugged in the formulas of Table 5.5 in order to obtain final numeric sensitivity indices. Table 5.6 presents the sensitivity indices of the parameters γ, q and β for different allocations of these parameters as well as ε_I and ε_H, wherefore the following assignments were used: $T_I = 2.214, T_H = 1.286, \iota = 0.2, M = 0.11$ and $K = 0.067$. It is assumed that $\varepsilon_C = 0$.

To obtain a numeric result, it if for simplicity determined that $q(t) = q_1$ and $\gamma(t) = c_1$ in the computations .

Table 5.6 Sensitivity indices of the parameters β, γ and q for different allocations of $\beta, \gamma, q, \varepsilon_I, \varepsilon_H$

β	γ	q	ε_I	ε_H	index of β	index of γ	index of q
0.0012	50	0.1	0.2	0.05	5.1753	0.00012321	−0.0069
0.00155	50	0.1	0.05	0.05	5.1753	0.00016043	−0.0089
0.00155	60	0.1	0.05	0.05	6.2103	0.00016043	−0.0107
0.00155	70	0.1	0.05	0.05	7.2454	0.00016043	−0.0125
0.00155	50	0.25	0.05	0.05	4.3127	0.00013369	−0.0089
0.00155	50	0.5	0.05	0.05	2.8751	0.00008913	−0.0089
0.0017	50	0.1	0.05	0.05	5.1753	0.00017596	−0.0098
0.0023	50	0.1	0.05	0.05	5.1753	0.0002306	−0.0132
0.00155	50	0.01	0.15	0.05	15.1383	0.00046929	−0.0261
0.00155	50	0.01	0.25	0.05	25.1013	0.00077814	−0.0432
0.00155	50	0.01	0.05	0.005	5.0009	0.00015503	−0.0086
0.00155	50	0.01	0.04	0.04	4.1402	0.00012835	−0.0071

Table 5.6 reveals that modifying β from 0.12 to 0.155, then further to 0.17 and 0.23 leads to an increase in γ of around 30%, then another 10% and 31%. This also results in slight changes in the size of q, which are 29%, another 10% and 35%, respectively. Changing the size of the parameter γ from 50 to 60 (70) creates an increase in the size of both β and q by 20% (40%). Increasing q from 0.1 to 0.15 (0.4) leads to a reduction in the size of both the indices of β and γ by around 17% (45%). At last, enhancing the size of the index of ε_I by 0.1 leads to a 2.925-fold (4.85-fold) increase in the size of β, while a reduction of ε_H by one tenths results in only a slight decrease in the size of the index of β. Reducing both ε_I and ε_H by 0.01 leads to a decrease of 20% in the index of β. Hence substantially, the great influence of the modification factors ε_I and ε_H as well as the counteracting effect of the size of q on the parameter β become obvious in this table.

Parameter Estimation in $MATLAB$ 6

In Chapter 6 of this thesis, the formation process the $MATLAB$ programs, which are used to obtain predictions of future compartment sizes on the basis of currently existing compartment size data in the form of time series forecasts, is explained. The presented SARS-CoV-2-fitted compartment model variations, which underlie the transitions between compartments expressed by the introduced systems of ODEs, can be characterized as dynamical models since they refer to real world processes changing over time.

In the model calibrations, classical methods of parameter estimation in systems of ODEs can be applied. The *nonlinear least squares* (NLS) method is explained in Subsection 6.1.1 and delivers an estimate for the parameter values by minimizing a discrete ℓ^2-error (*least squares*). In Subsection 6.1.2, on overview of the implementation procedure is given. Shortened to one sentence, this comprises the minimization of the discrete ℓ^2-error between the data that results from the numerical integration of a system of ODEs and evolving over multiple points in time, and accessible time series compartment size data, that are both based on the same initial compartment size data, by optimizing specific model parameters with the aid of the NLS method for the purpose of using the system of ODEs and optimized parameters to obtain forecasts of the time series progressions of future compartment size data.

Section 6.2 shows all parameter allocations and intervals that are set in the $MATLAB$ implementations. Intervals of the parameters shaping the trigonometric contact and quarantine rate are also given. Subsection 6.2.1 depicts the selected parameter values along with their definitions in a table for the $SIHCDR$ model

Supplementary Information The online version contains supplementary material available at (https://doi.org/10.1007/978-3-658-35932-4_6).

S. M. Treibert, *Mathematical Modelling and Nonstandard Schemes for the Corona Virus Pandemic,* BestMasters, https://doi.org/10.1007/978-3-658-35932-4_6

applied to data of the countries Germany and Sweden, and Subsection 6.2.2 does the same with the $SVID$ age group model applied to German data.

Compartment size predictions for the year 2021 are provided along with graphs of the time series data resulting from the $MATLAB$ implementations in Sections 6.3 and 6.4. In particular, the effects of the changes in parameters bounds of the transmission risk, the modification factors of the transmission risk, the quarantine rate as well as the contact rate are examined in the implementation of the $SIHCDR$ model in Section 6.3 to evaluate the impact of possible scenarios of intervention. The effect of variations in the vaccination rate is investigated in the $SVID$ age group model in Section 6.4 to analyse possible vaccination scenarios.

In Section 6.5, *nonstandard finite difference schemes* (NSFD) are introduced and utilized to replace the $MATLAB$ solver $ode45$ in certain implementations. The target is to guarantee the positivity of the solution and the correct long-time asymptomatic behaviour.

It is clear that multiple scenarios of distinct parameter allocations can be observed and analysed by calibrating the model parameters and generating plots depicting compartment size progressions. It is impossible to include all informative plots in this thesis. A selection of conclusive plots created by specific parameter allocations is incorporated in Sections 6.3, 6.4 and 6.5. In particular, the modification of the parameters β, q_1, c_1, ε_I and ε_H is focused on. The $MATLAB$ scripts that were created and used to make compartment size forecasts in this thesis can be found in Appendix B. It is emphasized at this point that the method of sensitivity analysis concerning the basic reproduction number R_0 is meaningful in regards of the effects of a parameter modification on the allocations of other model parameters necessary to attain certain future compartment size levels.

6.1   Problem Formulation and Implementation Process

During the implementation process, the ℓ^2-error minimization is pursued by making use of the NLS method. This method is explained in Subsection 6.1.1.

In Subsection 6.1.2, the process of implementation with reference to four distinct program scripts that interlock, meaning access each other, is described. It is also mentioned which time intervals and countries are used to equip the implementations with data.

6.1.1   The Nonlinear Least Squares Approach to Compartment Models

Let the term *measurement* describe a quantification of a specific state variable, and $D_i(t_j)$, $j \in \{1, ..., l\}$, $i \in \{1, ..., N\}$ be the measurement concerning the state variable x_i, that represents the size of the compartment $\mathcal{K}_i$, at some time point $t_j \in [t_0, t_l]$ in this context.

Let the measured data for the ith compartment and the jth point in time be saved in the vector

$$D_j^{(i)} := D_i(t_j).$$

Subsequently, a *measurement function* $\Phi(t)$ exists that maps a point in time $t_j \in [t_0, t_l]$ into a measured s-dimensional data set:

$$\Phi : \mathbb{R} \to \mathbb{R}^s, \Phi(t_j) := [D_j^{(1)}, ..., D_j^{(s)}]^\top \in \mathbb{R}^s$$

The complete set of measured data covering all compartments and all points in time $t \in [t_0, t_l]$ is saved as a matrix of the size $l \times s$, that can be transformed into a vector $\hat{\Phi}$ of the form

$$\hat{\Phi} := [\Phi(t_0), ..., \Phi(t_l)]^\top \in \mathbb{R}^n$$

with $n = l \cdot s$. In the sequel, the term "model-generated data" refers to the data obtained by integrating a system of ODEs $F(t, x, \vartheta)$ from t_0 to t_l for a given parameter vector ϑ and initial condition $x(t_0) = x_0$ stated in the initial value problem of Equation (5.1). At next, let $Y_i(t_j, \vartheta)$ be the program output data for the compartment $\mathcal{K}_i$, $i \in \{1, ..., s\}$ and the point in time t_j, $j \in \{1, ..., l\}$, which is abbreviated by

$$Y_j^{(i)} := Y_i(t_j, \vartheta).$$

Consequently, a model function $\mathcal{Y}(t, \vartheta)$ exists that maps a point in time t_j into a set of generated data for a given parameter vector ϑ:

$$\mathcal{Y} : \mathbb{R} \times \mathbb{R}^m \to \mathbb{R}^s, \mathcal{Y}(t_j, \vartheta) = [Y_j^{(1)}, ..., Y_j^{(s)}]^\top.$$

The complete set of model-provided data covering all s compartments and all points in time $t \in [t_0, t_l]$ is saved as a matrix of the size $l \times s$, that can be transformed into a vector $\hat{\mathcal{Y}}$ of the form

$$\hat{\mathcal{Y}}(\vartheta) := [\mathcal{Y}(t_0, \vartheta), ..., \mathcal{Y}(t_l, \vartheta)]^\top \in \mathbb{R}^n.$$

A model output data set $\hat{y}(\vartheta)$ obtained from the integration of a system of ODEs with specific initial conditions can be fit to a given time series data set $\hat{\Phi}$ as optimally as possible by optimizing the adjustable part of the model parameters ϑ. Therefore, let the adjustable part of the parameter vector ϑ, which is to be optimized, be $\vartheta_1 \in \mathbb{R}^{m_1}$, and the fixed part of ϑ be $\vartheta_2 \in \mathbb{R}^{m_2}$ with $m_1 + m_2 = m$. In the sequel, an entry of the vector $\hat{\Phi}$ is denoted by $\hat{\Phi}_k, k \in \{1, ..., n\}$ and an entry of the vector $\hat{y}(\vartheta)$ is denoted by $\hat{y}_k(\vartheta), k \in \{1, ..., n\}$.

A nonlinear optimization problem is a NLS problem if the objective function f has the form of so-called squared residuals. The target of a least squares minimization is to best fit certain data $\hat{y}(\vartheta) \in \mathbb{R}^n$ outputted from the integration of an ODE model to given reported or measured data $\hat{\Phi} \in \mathbb{R}^n$ by finding those model parameter values for the adjustable ϑ that minimize the least squares error between $\hat{y}(\vartheta)$ and $\hat{\Phi}$. Measurement of the fit between a model and a reported data point is performed by drawing on residuals.

A least squares method minimizes the sum of squared residuals in order to find optimal parameter values.

Data fitting and parameter estimation are the most relevant areas of application of least squares minimization.

In the case of parameter estimation for an epidemic forecast a system of ODEs of an enhanced SIR model is usually used as a model for the purpose of outputting compartment size data of s compartments at l different observed points in time. Letting s be the number of compartments of the underlying epidemic model which the model function $\mathcal{Y}$ is based on, the number of values comparable between measurements and model outputs is $n = l \cdot s$.

We assume the same l observation points in time for each of the s compartments $\mathcal{K}_1, ..., \mathcal{K}_s$.

Let the mentioned residuals be denoted by r, with $r : \mathbb{R}^m \rightarrow \mathbb{R}^n$, for which it holds

$$\forall k \in \{1, ..., n\} : r_k(\vartheta) := \hat{\Phi}_k - \hat{y}_k(\vartheta) \text{ with } \hat{y}(\vartheta) \in \mathbb{R}^n, \hat{\Phi} \in \mathbb{R}^n, \vartheta \in \mathbb{R}^m.$$

The objective function of the respective least squares problem is defined as

$$f : \mathbb{R}^m \rightarrow \mathbb{R}, f(\vartheta) = \sum_{k=1}^{n} r_k(\vartheta)^2 .$$

Following the above information, the goal of the established algorithms is the minimization of the discrete ℓ^2-error between reported and modelled compartment size data in order to make use of the parameter values minimizing this error for later model predictions. Mathematically expressed and using the Euclidean norm, the resulting *unconstrained optimization problem* has the objective function

$$\min_{\vartheta \in \mathbb{R}^m} \quad f(\vartheta) = \sum\nolimits_{k=1}^{n} r_k(\vartheta)^2 = ||r(\vartheta)||_2^2 = ||\hat{\Phi} - \hat{y}(\vartheta)||_2^2 = \sum\nolimits_{k=1}^{n} \left(\hat{\Phi}_k - \hat{y}(\vartheta) \right)^2.$$

The first and second order necessary and sufficient optimality conditions, which are written in various books and scripts dealing with numerical optimization as [258, pp. 37–39], apply to unconstrained optimization problems. It is generally difficult to globally solve NLS problems, but a local minimum can be found by iteratively solving the problem in order to linearise it at its current guess per iteration [258, p. 49]. To NLS problems, the *Gauss-Newton* (GN) method is often applied, which linearises the nonlinear function $r(\vartheta)$ at an iterate ϑ_z inside the ℓ^2-norm, such that the next iterate ϑ_{z+1} is obtained by solving a linear least squares problem of the form

$$\vartheta_{z+1} = \operatorname*{argmin}_{\vartheta} ||r(\vartheta_z) + J(\vartheta_z) \cdot (\vartheta - \vartheta_z)||_2^2,$$

where $J(\vartheta) = \frac{\delta r(\vartheta)}{\delta \vartheta}$ is the Jacobian matrix. Thus an iterative update of this method works in the following way, stated in [258, p. 49]:

$$\vartheta_{z+1} = \vartheta_z - \left(J^\top(\vartheta_z) \cdot J(\vartheta_z) \right)^{-1} J^\top(\vartheta_z) \cdot r(\vartheta_z).$$

The Gauss-Newton method works well for small residual problems. For a solution with perfect fit, a locally quadratic convergence rate is reached at the end of the iterates [258, p. 56].

In this thesis, the target is to find those adjustable model parameters $\vartheta_1 \in \mathbb{R}^{m_1}$ that yield a prediction $\hat{y}(\vartheta)$ as close as possible to the reported data points $\hat{\Phi}$ [258, p. 41]. In the implementations of this thesis, a built-in *MATLAB* solver, that is based on the so-called *interior-reflective Newton method*, is used instead of the GN method.

The term initial conditions describes the measured or reported data with respect to the first observed point in time t_0, that is $x_0 = \Phi(t_0)$. A point in time is a week in the following considerations and *MATLAB* implementations. Observed points in time refer to calendar weeks between the 10th calendar week in 2020 and the 7th calendar week in 2021. This time span is chosen because SARS-CoV-2 incidence sharply increased in Germany and Sweden from the beginning of March, and data regarding the numbers of newly confirmed infections, hospitalizations and deaths were available until mid-January at the time of implementation. However, not all of the corresponding weeks have been incorporated into all of the upcoming

implementations. More detailed information concerning the data usage is given in
Section 6.2.

6.1.2   The Implementation Process

The implementation process is subdivided into four program scripts that intertwine.
The whole process is applied to the $SIHCDR$ or $SVID$ model and applicable
to other compartment models, too, whereby the $SIHCDR$ model is applied to
compartment size and parameter data of Germany or Sweden and the $SVID$ age
group model to data of Germany. Data sources are stated in the introduction of
Section 6.3. The system of ODEs of the underlying model is implemented in the
first program script. All of the model parameters in the parameter vector ϑ are
passed to this script by the second program script for the purpose of obtaining the
modelled data $Y(t, \vartheta)$ mentioned in Subsection 6.1.1 by applying an integrator to
the system of ODEs of the first script. The $MATLAB$ solver $ode45$ is used here
as a first option, the method of *nonstandard finite difference* (NSFD) schemes is
realized as an alternative.

In the second script, the data $\hat{\Phi}(t)$ representing the sizes of the compartments
of the selected model for l different point in time, are initialized and the discrete
ℓ^2-error between the reported compartment size data Φ and modelled compartment
size data $\hat{Y}(t, \vartheta)$ is computed. The parameters ϑ_2 that are not estimated during the
process are initialized here, whereas the parameters ϑ_1 are received from the third
program script.

In the third script, a NLS minimization as described in Subsection 6.1.1 is performed by using the built-in $MATLAB$ function $lsqnonlin$, that is applied to the
least squares error computed using the second program script. Lower and upper
bounds of the optimized model parameters ϑ_1 are selected and passed to the function $lsqnonlin$ as well. An alternative would be the built-in $MATLAB$ function
$fminsearch$, which is based on the *Nelder Mead algorithm*, that is also called
Downhill Simplex method. In Appendix A, the idea and operating of the *trust-
region method*, which is used by the $MATLAB$ function $lsqnonlin$, is explained.

In the fourth program script, parameter values for ϑ_2, which are used to model
and plot compartment sizes over future points in time, are selected. They can differ
from the values for ϑ_2 utilized for the computation of the discrete ℓ^2-error in the
second script. For instance, the values for the non-estimated parameters can fit real
country-specific values of the first wave of the pandemic in the error computation,
but the country-specific real values of the second wave of the pandemic in the

development of the compartment sizes over time. The initial values of the sizes of all compartments of the used model have to be selected as well.

An implementation process is realized per country for both the underlying data of the first and second wave of the pandemic. The compartment size data of the calendar weeks 10 until 38 (called the "first wave") in 2020 are used for the prediction of the second wave, and the compartment size data of the calendar weeks 10 in 2020 until 24 in 2021 are used for the prediction of a possible third wave that occurs over the autumn/winter 2021/2022. The calendar weeks 39 in 2020 until 24 in 2021 are referred to as the "second wave". Consequently, the two time periods November 2020 to January 2021 and spring 2021, when peaks occurred in Europe, are summarized as one "wave". The reason for that is the fact that the implementations of this thesis were finalized at beginning of March 2021, when compartment size data concerning the end of March, April etc. were not yet available.

By changing the model parameters in the third or fourth script, the courses and local extrema of the resulting compartment functions are varied. Therefore, the whole implementation can be adjusted to distinct scenarios like a modified transmission rate due to mutations or vaccination programs, or modified quarantine and contact rates owing to implemented intervention measures. Plots of compartment evolution over time with applied calibration are explained and shown in Sections 6.2, 6.3 and 6.4.

6.2 Parameter Definitions and Bounds

A part of the transmission and transition parameters mentioned in Sections 4.2 and 4.3 is reported, another part is computable from certain other reported parameters, and the rest is estimated. In this section, it is explained which parameters are fixed to a value, kept variable in an interval, are estimated or can be computed on the basis of other parameters with regard to the implemented compartment models.

6.2.1 Parameters in the $SIHCDR$ Model

In the following $MATLAB$ implementations of the $SIHCDR$ model, the transmission risk β, the quarantine rate $q(t)$, the transmission modification factors ε_I and ε_H, the fraction of individuals admitted to intensive care ι as well as the CFR for ICU patients M_ι are the parameters that are optimized. Moreover, the transmission rate $\theta_I(t)$ is defined as in Subsection 5.2.2 and the transition parameters $\eta^I, \omega_2{}^I, \lambda_1, \xi, \omega_3, \lambda_2$ and ω_4 as explained in Chapter 4. Reported values are used for the allocation of the parameters T_I, T_H, T_t, and the values for M, M_H and K are computed from reported data.

The contact rate $\gamma(t)$ and the quarantine rate $q(t)$ are defined as trigonometric functions. A realistic number of contacts per person per day is assumed to lie between 11 and 20 in European countries [259]. Thus a person is assumed to normally have between 77 and 140 relevant contacts per week. This influences the contact rates used in the program runs. As mentioned in Subsection 6.1.1, compartment size and parameter data concerning the first wave of the pandemic are used to model the second wave, and such data with respect to the average of both the first and second wave are made use of in order to model a probable third wave. Case-fatality rate and case-hospitalization rate parameters concerning the calendar weeks 10 to 38 in 2020 are named M_1, M_{H1}, K_1, and those referring to the calendar weeks from week 39 in 2020 are named M_2, M_{H2}, K_2.

Figure 6.1 depicts examples of contact rates applied in the implementations. Here, the parameter c_2 is fixed to a value for each of the pairs first wave and Germany, second wave and Germany, first wave and Sweden as well as second wave and Sweden. However, it is varied within the interval presented in Table 6.1 per pair in the implementations in order to adapt the algorithm to COVID-19 provisions. The separate consideration of the first and second wave enables an even more realistic rate modelling. The amplitude of the respective contact rate is selected as 40 in the first and 20 in the second wave for Germany and Sweden. The parameter c_0 is set to the value 20 for both Germany and Sweden. The left diagram of Figure 6.1 shows the function $(40 - 20) \cdot cos\left(\frac{\pi}{20} \cdot (t - 40)\right) + 60$ for Germany and $(40 - 20) \cdot cos\left(\frac{\pi}{20} \cdot (t - 40)\right) + 120$ for Sweden in the interval [0,15] on the x-axis (January until early April), whereas the right picture illustrates the contact rate function $(30-20) \cdot cos\left(\frac{\pi}{20} \cdot (t-13)\right) + 70$ for Germany and $(30-20) \cdot cos\left(\frac{\pi}{20} \cdot (t-13)\right) + 90$ for Sweden in the interval [30,70] on the x-axis (end of July until March). The parameter c_1 is selected as generally higher for Sweden than Germany in both cases due to the Swedish special path during the pandemic.

The quarantine rate function used in the implementation for both waves has the form $q(t) = q_1 \cdot cos\left(\frac{\pi}{20} \cdot t\right) + q_1$, so $z_2 = 0$, in order that a minimum is reached in the calendar week 30 and a maximum is attained in week 50 in 2020. This means that the maximal quarantine ratio is reached in the same week as the minimal contact ratio is reached, namely in the second week of December 2020.

The unit of the contact ratio on the x-axis in Figure 6.1 is $\frac{contacts}{week}$.

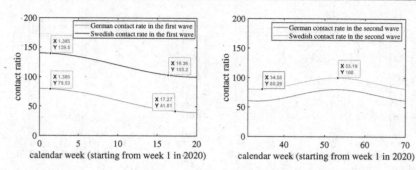

Figure 6.1 The trigonometric contact rate $(c_2 - c_0) \cdot cos\left(\frac{\pi}{20} \cdot (t - z_1)\right) + c_1$ for Germany and Sweden in the first wave (left picture) and second wave (right picture) for $c_2 = 40$ (first wave), $c_2 = 30$ (second wave), $c_0 = 20$, $c_1 = 60$ (Germany, first wave), $c_1 = 70$ (Germany, second wave), $c_1 = 120$ (Sweden, first wave), $c_1 = 90$ (Sweden, second wave) and $z_1 = 0$ (first wave), $z_1 = 13$ (second wave)

It is reasonable that the contact rate significantly decreases between February and the beginning of May 2020, as shown in the left diagram of Figure 6.1, due to the fact that the population was put on the alert in February and first interventions came into force in Europe in mid-March 2020. It is also realistic that the level of the number of contacts per person is relatively low over the course of the second wave, which was stronger than the first one, with a little elevation and so a local maximum around Christmas. Thus the right diagram in Figure 6.1 illustrates that a local maximum (80 for Germany, 100 for Sweden here) is reached in the calendar week 53, and local minima are observable in the calendar weeks 34 in 2020 (mid-August) and 16 in 2021 (mid-April).

The transmission rate $\theta_I(t)$ as well as the rate $\theta_{S_q}(t)$ are exemplarily depicted in Figure 6.2 for the values $q_1 = 0.002$, $c_0 = 20$, $c_1 = 50$, $c_2 = 30$, $z_1 = 13$, $z_2 = 0$ and a random allocation of $\beta = 0.003$. The transmission rates used in the later implementations have the form of $\theta_I(t)$ to model the transmission between the compartments S and I.

The unit on the y-axis is $\frac{effected\ transmissions}{week}$ in the figure below. It has to be kept in mind that the rate $\theta_I(t)$ is multiplied by the current size of the susceptible compartment in the first two equations of the system of ODEs of the $SIHCDR$ model (cf. Table 5.1) to obtain the number of susceptibles infected per individual in the class I per unit of time [200, p. 10]. The force of transmission equals the transmission risk here and can be interpreted as the probability of transmission per contact.

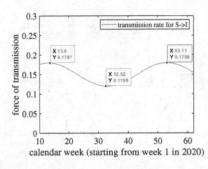
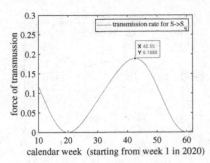

Figure 6.2 The transmission rates $\theta_I(t)$ (left picture) and $\theta_{S_q}(t)$ (right picture) for a maximum quarantine ratio of 0.4%, $\beta = 0.003$, and a minimum (maximum) number of contacts of 40 (60), shown in a progress over 62 weeks

The rate of transition from the susceptible to the infected class of infection increases until a local maximum of around 0.18 is reached between the end of March and the beginning of April, then decreases to a minimum of around 0.12 (week 32, beginning of August) and finally increases again. Hence, the number of newly infected individuals at time t is larger than 12% of the current number of infectious individuals in I multiplied by a current normalized number of susceptibles in the system at all time instants t in this example.

The rate of the transition from the compartment S to S_q fluctuates in the interval $[0, 0.19]$, reaching a local minimum of zero in May and maximum in late October. Thus the number of susceptibles put in quarantine and influenced by NPIs at time t is at most 18% of the number of current infectious cases in I at time t multiplied by a normalized number of susceptibles for all time instants t here. The progressions of the two graphs in Figure 6.1 are reasonable in the way that the first wave of infection reached Germany in March 2020 such that the number of daily newly confirmed infections reached a peak on April 1st (calendar week 14) [12]. Then the number of weekly newly confirmed infections sank until the end of the summer 2020, and increased in autumn/winter until a new maximal value was attained around the turn of the year. Simultaneously, the strong growth in quarantine imposition and NPI realization in the end of October is realistic, since the second wave reached Germany in the beginning of October and several state interventions were authorized from then.

Table 6.1 presents all of the parameters used in the implementations of reduced form of the $SIHCDR$ model for the two countries Sweden and Germany. It contains all parameter definitions, parameter calculation formulas as well as parameter values

Table 6.1 Selected parameter values and definitions for the $SIHCDR$ Model for Germany and Sweden

Parameter	Parameter Definition	Sourcing	Germany	Sweden	Unit
N	population size	Reported	83, 100, 000 [260]	10, 380, 245 [261]	inhabitants
L	life expectancy in years	Reported	81 [262]	81.85 [263]	years
μ	weekly natural death rate	$\frac{1}{L \cdot 52}$	0.0002374	0.0002350	$\frac{1}{years}$
β	transmission risk parameter	Estimated	[0.0011, 0.00236]	[0.00157, 0.0029]	$\frac{infections}{contact}$
c_0	first contact rate parameter	exemplary	20	20	$\frac{contacts}{week}$
c_1	second contact rate parameter	exemplary	[50,108]	[70,110]	$\frac{contacts}{week}$
c_2	amplitude of the contact rate	exemplary	30 or 40	30 or 40	$\frac{contacts}{week}$
z_1	shift of $\gamma(t)$ on the x-axis	exemplary	different	different	–
$\gamma(t)$	time-dependent contact rate	$(c_2 - c_0)cos\left(\frac{\pi}{20}(t - z_1)\right) + c_1$	–	–	$\frac{contacts}{week}$
q_1	determines max. quarantine ratio	Estimated	[0,0.5]	[0,0.5]	–
$q(t)$	time-dependent case quarantine rate	$q_1 cos\left(\frac{\pi}{20}t\right) + q_1$	–	–	–
ε_I	modification of the transmission rik for I	Estimated	[0.1,0.5]	[0.1,0.5]	–

(continued)

Table 6.1 (continued)

Parameter	Parameter Definition	Sourcing	Germany	Sweden	Unit
ε_H	modification of the transmission risk for H	Estimated	[0.05,0.0625]	[0.05,0.0625]	–
$\theta_I(t)$	transmission rate for $S \rightarrow I$	$\beta \cdot \gamma(t) \cdot (1 - q(t))$	–	–	$\frac{infections}{week}$
$\Theta_I(t)$	rate for transition $S \rightarrow I$	$\beta \cdot \gamma(t) \cdot (1 - q(t)) \cdot X_I(t)$	–	–	–
$\mathcal{L}$	length of latency period	Reported	2.5/7 [46]	2.5/7	weeks
$\mathcal{T}^I$	length of incubation period	Reported	5.5/7 [10]	5.5/7	weeks
τ	average time spent as an asymptomatic infectious	$\mathcal{T}^I - \mathcal{L}$	0.423	0.423	weeks
T	time from symptom start until recovery	Reported	1.429 [10]	1.429	weeks
T_I	length of contagious period	Reported	2.214 [10]	2.214	weeks
T_H	time from hospitalization until recovery	Reported	1.286 [10]	1.286	weeks
T_ι	time from ICU admission until recovery	Reported	1 [10]	1	weeks

(continued)

Table 6.1 (continued)

Parameter	Parameter Definition	Sourcing	Germany	Sweden	Unit
M	CFR in CW 10/2020-05/2021	Reported	0.024 [264]	0.041 [265]	–
M_1	CFR in CW 10/2020-33/2020	Reported	0.031 [264]	0.076 [265]	–
M_2	CFR in CW 34/2020-05/2021	Reported	0.019 [264]	0.011 [265]	–
M_H	CFR for H in CW 10/2020-05/2021	$\dfrac{M\cdot I_{ges}-0.6\cdot C_{ges}}{H_{ges}}$	0.22	0.37	
M_{H1}	CFR for H in CW 10/2020-33/2020	$\dfrac{M_1\cdot I_1-0.6\cdot C_1}{H_1}$	0.066	0.12	
M_{H2}	CFR for H in CW 34/2020-05/2021	$\dfrac{M_2\cdot I_2-0.6\cdot C_2}{H_2}$	0.18	0.11	
M_ι	CFR for C in CW 10/2020-05/2021	**Estimated**	[0.4,0.7]	[0.4,0.7]	–
K	CHR in CW 10/2020-05/2021	Reported	0.11 [264]	0.17 [266]	–
K_1	CHR in CW 10/2020-33/2020	Reported	0.15 [264]	0.30 [266]	–
K_2	CHR in CW 34/2020-05/2021	Reported	0.067 [264]	0.061 [266]	–
ι	fraction of hospitalized who transit to ICU	**Estimated**	[0.13,0.178]	[0.17,0.26]	–

(continued)

Table 6.1 (continued)

Parameter	Parameter Definition	Sourcing	Germany	Sweden	Unit
η^I	case-hospitalization rate for I	$\frac{1-\mu T_I}{T_I} \cdot K$	–	–	–
ω_2^I	recovery rate for I	$\frac{1-\mu T_I}{T_I} \cdot (1-K)$		–	–
λ_1	disease-induced case-fatality rate for H	$\frac{1-\mu T_H}{T_H} \cdot M_H$	–	–	–
ξ	rate for transfer to intensive care	$\frac{1-\mu T_H}{T_H} \cdot \iota$			–
ω_3	recovery rate for H	$\frac{1-\mu T_H}{T_H} \cdot (1-M_H-\iota)$	–	–	–
λ_2	disease-induced case-fatality rate for C	$\frac{1-\mu T_\iota}{T_\iota} \cdot M_\iota$	–	–	–
ω_4	recovery rate for C	$\frac{1-\mu T_\iota}{T_\iota} \cdot (1-M_\iota)$	–	–	–

used for the adoption of the system of ODEs to the countries Germany and Sweden. The case-hospitalization are is abbreviated with CHR here. No units are indicated for transition rates denoting fractions of population groups.

Exemplary intervals are given for the three contact rate parameters. This means that they are not reported and not estimated by the algorithms, but fixed in reasonable intervals based on preliminary considerations.

Furthermore, the values for the different time periods to recovery T, T_I, T_H and T_ι are assumed to not relevantly differ between Germany and Sweden, and so are obtained from the Robert-Koch Institute Germany and adopted for Sweden, too. The average time of a COVID-19-induced hospital stay is assumed to last 9 days (1.286 weeks), intensive care treatment to last 7 days since the total time of hospital stay of patients admitted to intensive care is assumed to last 16 days, time from symptom development until recovery to last 10 days (1.429 weeks), and length of contagious period to last 15 days (2.214 weeks) [10].

The values for the rate ι were computed as 0.24 (0.19) for the combination first wave and Germany (Sweden), 0.18 (0.24) for the combination second wave and Germany (Sweden), and 0.20 (0.14) for both waves and Germany (Sweden) from the underlying data for the compartment sizes of the hospitalized and ICU compartment. It should be stressed again that recovery is identified with the loss of the ability to infect others rather than the complete disappearance of symptoms here.

With respect to data search, the time until hospital or ICU dismissal is used as an indicator for this. The case-fatality rate as well as case-hospitalization rate for certain the calendar weeks of 2020/21 are not directly obtained from a data source, but computed on the basis of reliable sources named in the table. It is assumed that no leaky vaccine or infectious compartments E_V, A_V, I_V are involved in the used model. The following function $X_I(t)$ is defined in order to be included in the incidence rate:

$$X_I(t) := \varepsilon_I \cdot I(t) + \varepsilon_H \cdot H(t) + \varepsilon_C \cdot C(t), \tag{6.1}$$

Individuals in the compartment C are not assumed to be able to be infectious to susceptible in the implementations. For this reason we set $\varepsilon_C = 0$. Letting l be the number of points in time for which compartment size data is acquired, an incidence rate of the form

$$\theta_I(t) \cdot X_I(t) \cdot \frac{S(t)}{\frac{S(t)}{l} + I(t) + H(t) + C(t) + R(t)} \tag{6.2}$$

is applied in the implementations.

6.2.2   Parameters in the $SVID$ Age Group Model

The implemented age group model substantially differs from the $SIHCDR$ model in the way that the infected compartments I, H and C are pooled as one infected class I, an all-or-nothing vaccinated compartment V is included, three age groups are regarded such that every compartment is divided into three (sub-)compartments, the contact and quarantine rates of the age groups are not implemented as time-dependent functions but constant values that can however be modified between different program runs, and rates describing the contacts between the distinct age groups are needed. It is obvious that the individuals in the infected compartments I_1, I_2 and I_3 can be infectious, symptomatic or even hospitalized but do not have to be. An age-group and time-dependent vaccination rate is introduced. It should be mentioned that the compartment R is omitted here since no reliable data related to age groups are available. Thus the individuals in the compartments I_1, I_2 and I_3 recover by leaving the system.

Table 6.2 provides the parameter definitions and values utilized for the implementation of the $SVID$ age group model based on the data for Germany. It has the same structure as Table 6.1, but depicts parameter allocations with reference to the age groups of 0–19-, 20-59- and over 60-year-old people instead of the countries Germany and Sweden. Data were available for the calendar weeks 10 in 2020 until 7 in 2021. In the implementation, placeholder data were used for the calendar weeks 8 to 26 in 2021.

As it can be seen from Table 6.2, the parameters $q_1, q_2, q_3, \epsilon_1, \epsilon_2, \epsilon_3$ and β are estimated. The contact ratios are selected as follows:

$$\gamma_{11} = 60, \gamma_{22} = 80, \gamma_{33} = 65, \gamma_{12} = \frac{7}{6} \cdot \gamma_{11} = 70, \gamma_{13} = \frac{5}{6} \cdot \gamma_{11} = 50, \gamma_{23} = \frac{5}{4} \cdot \gamma_{11} = 75 \tag{6.3}$$

such that

$$\gamma_{13} < \gamma_{11} < \gamma_{33} < \gamma_{12} < \gamma_{23} < \gamma_{22}. \tag{6.4}$$

The vaccination rates are modelled as exponentially increasing functions of the form $v(t) = v_A \cdot e^{\frac{t}{v_B}}$ for all age groups in order that an enlarged value of v_B diminishes the slope of $v(t)$. The initial vaccination ratio v_A is computed as the weekwise number of second vaccinations in the respective age group in Germany divided by the total size of the German population in this age group. The first secondary vaccination within the German population was officially received on January 5th 2021 [269]. The used data obtained from the vaccination monitoring program of the RKI were

Table 6.2 Selected parameter values and definitions for the $SVID$ Age Group Model for Germany

Parameter	Parameter Definition	Sourcing	0-19 years	20-59 years	60+ years	Unit
μ	weekly natural death rate	Reported	0.000001045 [267]	0.0000196 [267]	0.000198 [267]	$\frac{1}{years}$
β	transmission risk	Estimated	[0.025,0.027]	[0.025,0.027]	[0.025,0.027]	$\frac{infections}{contact}$
q	quarantine rate	Estimated	0.2	0.3	0.1	–
ε	modification of the transmission risk	Estimated	0.15	0.15	0.15	–
T_I	length of contagious period	Reported	$\frac{1}{2} \cdot 2.214$	2.214 [10]	$\frac{5}{2} \cdot 2.214$	weeks
M	CFR in CW 10/2020-07/2021	Reported	0.000373 [268]	0.000807 [268]	0.024 [268]	–
v_A	initial vaccination ratio	Reported	0.00004125 [269]	0.0120 [269]	0.0236 [269]	–
v_B	inverse exponent of vaccination rate	Reported	[8,13]	[18,28]	[24,33]	–
$v(t)$	vaccination rate	$v_A \cdot e^{\frac{t}{v_B}}$				–
ω	recovery rate	$\frac{1-\mu T_I}{T_I} \cdot (1-M)$				–
λ	disease-induced case-fatality rate	$\frac{1-\mu T_I}{T_I} \cdot M$				–

available up to February 21st 2021. It is assumed for simplicity that 70% of the second vaccinations in Germany were received by people of 60 years or older as of this date, 25% by the age group of 20 to 59-year-old people, and the remaining 5% by individuals between 0 and 19 years. As an additional information, the German Federal Statistical Office stated that 29% of the German population were scheduled for vaccination owing to their age as they were 60 years or older as of mid-February 2021 [270].

If the values $v_B^{(0-19)} = 11$, $v_B^{(20-59)} = 25$ and $v_B^{(60+)} = 35$ are exemplarily selected, around two thirds of the over 60-year-old people are vaccinated 85 weeks, 20-59-year-old people 100 weeks and 0-19-year-old people 110 weeks after the start of the vaccination program. If the values $v_B^{(0-19)} = 10$ (9 or 8), $v_B^{(20-59)} = 23$ (21 or 19) and $v_B^{(60+)} = 32$ (29 or 26) are selected instead, around two thirds of the over 60-year-old people are vaccinated after 76 (69 or 62) weeks, 20–59-year-old people after 93 (84 or 76) weeks and 0–19-year-old people after 97 (87 or 77) weeks. Choices of v_B similar to these exemplary ones are justifiable with respect to the observed time span of 60 weeks starting from the calendar week 26 in 2021, since it is assumed that it takes at least one year to vaccinate all German inhabitants who want to get vaccinated against COVID-19. Apart from this, a small portion of the German population does not want to receive the vaccination for different reasons in the long term according to the current state of knowledge.

The CFR for each age group is computed as the CFR of the whole German population provided by the RKI [264] multiplied by the fraction of total COVID-19-confirmed deaths of the respective age group, of all COVID-19-confirmed deaths reported to the RKI as of February 16th 2021 [268]. The length of the contagious period T_I of the under 20-year-old people is assumed to be half as long as for the 20–59-year-old individuals while the infectious period of the over-60-year old people is assumed to last two and a half times as long as the one of the 20–59-year-old people.

6.3   Compartment Size Predictions with the $SIHCDR$ Model

Compartment size data were obtained from the online source Statista.de [264] (number of newly confirmed infected, hospitalized and deceased individuals), Intensivregister [271] (number of new COVID-19 intensive care unit patients) and Worldometers [272] (number of newly recovered people) for Germany, as well as Statista.de [273, 274] (number of newly confirmed infected or deceased individuals) and the PHA [275] (number of newly hospitalized and new COVID-19 intensive care unit patients) for Sweden. Placeholder data are used as input compartment size data for the calendar weeks 6 until 24 in 2021. Here, it is assumed for both Germany and Sweden that incidence rose in the 3 weeks after the carnival week of 2021 and decreased again afterwards, since it is reasonable that the number of new infections increases due to carnival celebrations, which was also observed in 2020.

Regarding the number of hospitalized or ICU patients, the number of individuals newly admitted at time instant t was used as the size of the class $H(t)$ or $C(t)$, respectively. Since it is a common assumption that individuals remain infected for more than a week and $T_I > T > T_H > T_l$, it seems sensible to use more than 1 time (here 1.5 times is applied) the number of the people newly infected at time instant t as the size of the compartment $I(t)$. The weekwise compartment size $R(t)$ and $D(t)$ was computed as the cumulated number of weekly recovered or deceased individuals, respectively.

The weekwise numbers of recovered people in Sweden were not publicly available such that they were computed as 90% of the correspondent number of newly confirmed infected cases of 2 weeks before. In general, it is impossible to obtain or compute the number of weekly newly recovered individuals completely accurately because it is uncertain when every single infected person loses his or her contagiousness. Moreover, the fact that only positively tested (which means confirmed SARS-CoV-2 cases) are included, as reliable data is unavailable and would be unreliable with respect to unconfirmed cases, has to be kept in mind in order to be able to correctly assess the dimension of the compartment size data obtained from the data sources and implementations.

Generally, the program scripts described in Subsection 6.1.2 are used in Section 6.3. In Subsection 6.3.1, compartment size data of the calendar weeks 10 to 38 in 2020, which describe the first wave, are utilized to predict the sizes of the compartments S, I, H, C, D and R in the second wave as exactly as possible. In Subsection 6.3.2, the compartment size data of the calendar weeks 10 in 2020 to 24 in 2021, including placeholder data for the weeks 6 to 24 in 2021, are utilized to predict the sizes of the compartments S, I, H, C, D and R in a possible third wave.

It has to be mentioned that all estimated parameters except for q_1, that was optimized to its upper bound, were minimized to their lower bounds in the following scenarios, since the bounding intervals were selected tight for the purpose of obtaining as realistic and contrastable courses of compartment size development as possible.

Furthermore, the unit for the compartment size denoted on the y-axis of each graph in the following sections is "individuals", and the x-axis represents the passed time in weeks.

6.3.1   Prediction of the Second Wave

In the implementations of the first wave to predict the second wave, the parameters K_1 and M_{H1} were utilized for the error minimization, and K_2 and M_{H2} for the predictions. The parameter c_1 was exemplarily fixed to the value 120 for Sweden and 60 for Germany here. Table 6.3 depicts the reported sizes of the infected compartments concerning the calendar weeks 50 in 2020 to 2 in 2021, which are the weeks in which maxima were reached for the compartments I, H, C, D and R in Germany and Sweden. As mentioned above, the number of individuals actually situated in the compartment I was multiplied by 1.5 here.

Table 6.3 Compartment size data for the infected compartments, for Germany and Sweden, and the calendar weeks 50 in 2020 to 2 in 2021

Calendar Week	Germany, I	Germany, H	Germany, C	Sweden, I	Sweden, H	Sweden, C
50 in 2020	234,743	10,505	1,719	65,093	1,989	254
51 in 2020	262,608	11,564	1,889	71,738	2,217	283
52 in 2020	209,148	10,243	1,982	58,631	2,387	311
53 in 2020	185,058	10,086	1,905	61,760	2,549	348
1 in 2021	218,552	10,043	1,773	55,700	2,553	372
2 in 2021	178,850	8,781	1,656	39,666	2,328	352

The aim of the prediction of the second wave is to show which assignments of the adjustable model parameters closely yield the high points observable in the reported data of Table 6.3. This can be evaluated best by plotting the compartment size data provided by the fourth mentioned $MATLAB$ program.

Primarily, the parameter β was varied. The results are depicted in Figure 6.3 for the infected compartment in Sweden, Figure 6.4 for the infected compartment in Germany, Figure 6.5 for the hospitalized compartment in Sweden, and Figure 6.6 for the hospitalized compartment in Germany.

For both Germany and Sweden, the upper bound of the value β was set to 0.002. The lower bound of β was varied in the interval [0.00157,0.00168] for Sweden and [0.00225,0.00236] for Germany. For all values of β and both countries, the bounds of other estimated parameters were initially set to $\iota \in [0.17, 0.4]$, $\varepsilon_H \in [0.05, 0.25]$, $\varepsilon_I \in [0.2, 0.5]$, $M_\iota \in [0.5, 0.7]$ and $q_1 \in [0.01, 0.1]$. With these initial assignments, the variation of β was the only factor influencing the differences in the outputted plotted curves per country. The values initially used for K_1, K_2 and M_{H1}, M_{H2} are given in Table 6.1.

The parameter ε_I was optimized to the value 0.2 and ε_H to the value 0.05 i.e. to their respective lower bound in all of the associated program runs. This means that the transmission risk emerging from confirmed and isolated infected cases (asymptomatic or symptomatic) is 20% and the transmission risk emerging from hospitalized patients compared is 5% of the transmission risk emerging from unisolated symptomatic infectious individuals here.

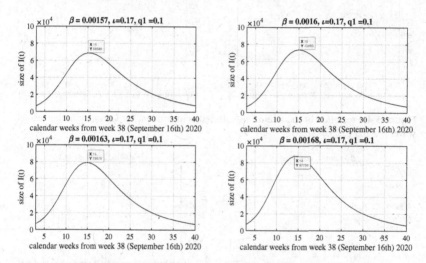

Figure 6.3 Prediction of the size of the infected compartment for Sweden in the second wave, with $\beta \in [0.00157, 0.00168]$

Variation of β

The upper left graph in Figure 6.3 demonstrates that the assignment $\beta = 0.00157$ produces a peak of 68, 680 people in the infected compartment for Sweden, whereas the other graphs in this figure show that the peak becomes 7.57% (15.28%) larger if β is enlarged by $3 \cdot 10^{-4}$ ($6 \cdot 10^{-4}$). Regarding these three allocations of β, the maximum is reached in the calendar week 1 in 2021 for Sweden. In contrast to this, a maximum of 87, 750 infected individuals is attained in the last calendar week in 2020 if the parameter β is assigned to the value 0.00168. The reported maximal size of the infected compartment at the turn of the year 2020/2021 in the second

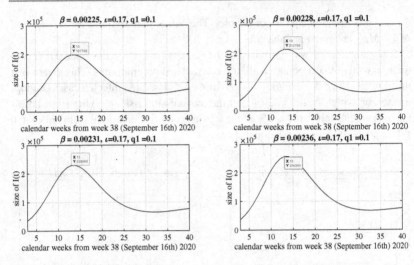

Figure 6.4 Prediction of the size of the infected compartment for Germany in the second wave, $\beta \in [0.00225, 0.00236]$

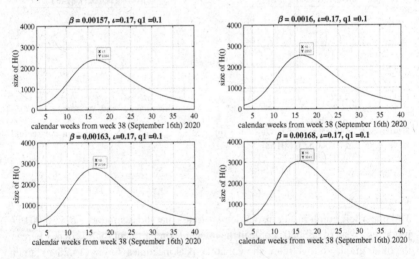

Figure 6.5 Prediction of the size of the hospitalized compartment for Sweden in the second wave, with $K_2 = 0.061$, with $\beta \in [0.00157, 0.00168]$

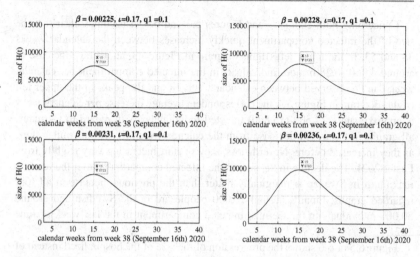

Figure 6.6 Prediction of the size of the hospitalized compartment for Germany in the second wave, with $K_2 = 0.067$, $\beta \in [0.00225, 0.00236]$

wave is 71,738 people in Sweden in the calendar week 51 (cf. Table 6.3), such that a value of β between 0.001557 and 0.00160 seems most realistic here. The peak attainment after 14 or 15 regarded calendar weeks in Figure 6.3 is realistic since a reported lasting decline in incidence is only notable from the calendar week 1 in 2021.

In Figure 6.4, it can be observed that the assignment $\beta = 0.00225$ leads to 197,700 individuals in the infected class in Germany in the calendar week 51 in 2020, which is the Christmas week. Enlarging the value of β by $3 \cdot 10^{-5}$ ($6 \cdot 10^{-5}$, $1.1 \cdot 10^{-5}$) results in a 7.6% (15.6%, 29.6%) larger $I(13)$. Thus the maximal number of people in the infected compartment is reached 7-14 days earlier in Germany than Sweden. Since a maximum of 262,608 infected people was reached in Germany at the turn of the year 2020/2021 according to Table 6.3 the assignment $\beta = 0.00236$ seems most realistic here. It may not be left out of sight that the number of newly infected individuals was multiplied by 1.5 concerning Table 6.3 in order to obtain a more realistic value of the number of individuals in the infected compartment per unit of time. The factor 1.5 is certainly not true in reality for all regarded weeks. Moreover, the dark figure may not be forgotten when analysing any graph of compartment size progression in a SARS-CoV-2 scenario.

Concerning both countries and Figures 6.3 and 6.4, it becomes clear that the size of the infected compartment quickly increases between the calendar weeks 38 and 53. In the upper left (right) diagram in Figure 6.3, an average increase of around 4,860 (6,280) more individuals in the infected compartment per calendar week can be observed between calendar week 38 and the peak. In the upper left (right) diagram in Figure 6.4, the correspondent average increase per calendar week is approximately 12,500 (15,800) infected individuals. With regard to Germany, all graphs in Figure 6.4 decrease from the calendar week 51 with a similar slope as they increased before. Nevertheless, they re-start increasing very slightly from Easter 2021. The decline in the size of the infected compartment after the reached maximum in Sweden is marginally flatter than the previous increase in all four regarded graphs. Finally, around 10,000 people are in the Swedish and around 80,000 individuals in the German infected compartment in the last week of June 2021.

Figure 6.5 or 6.6 shows the progression of the size of the hospitalized instead of infected class for Sweden or Germany, respectively. The parameter allocations are the same as in Figures 6.3 and 6.4. It can be seen that the courses of all four graphs have very similar slopes as the country-specific graphs in Figures 6.3 and 6.4. If β is set to the value 0.00157 for Sweden as to be seen in the upper left diagram, a peak of 2,384 hospitalized individuals is attained in the calendar week 3 in 2021. The peak is shifted to the calendar week 2 if β is increased to 0.00160, 0.00163 or 0.00168. All of the time, a larger β results in a larger peak. According to Table 6.3, a maximum of 2,553 Swedish COVID-19 hospitalizations was achieved in the calendar week 1 in 2021. Therefore, the assignment $\beta = 0.00160$ is most realistic with respect to the allocations of the other model parameters if the height of the peak is used as the basis of comparison. It has to be added that an exemplary modification of the parameter z_1 from 40 to 25 (−10) would lead to a shift in the calendar week in which the peak occurs from week 16 to 25 (27).

Relating to the upper left graph in Figure 6.6, where $\beta = 0.00225$ is applied, 7,561 hospitalizations can be observed in the calendar week 53 in 2020, whereas 9,763 hospitalizations in the same week are the result of the assignment $\beta = 0.00236$, which can be seen in the lower right graph. The increases in the reached maxima due to raised values of β in the last three graphs compared to the first graph in Figure 6.6 have the same percentage sizes as the growths in the peaks in Figure 6.4 owing to the same increases in β.

Variation of β and changed K_2, Germany
Figure 6.7 illustrates the change in the size of the compartment H if K_2 is set to the value 0.077 instead of 0.067 to attain a more realistic result for the maximally

reached number of hospitalized individuals in Germany in the second wave with a fixed lower bound of β.

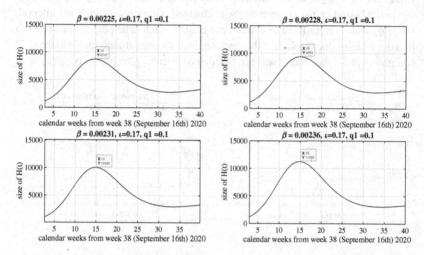

Figure 6.7 Prediction of the size of the hospitalized compartment for Germany in the second wave, with $K_2 = 0.077$, $\beta \in [0.00225, 0.00236]$

In the scenario shown in Figure 6.7, a larger K_2 leads to higher peaks in the number of hospitalizations compared to Figure 6.6, which has been expectable. The number of German hospitalizations is larger by 1,186 than in Figure 6.6 for $\beta = 0.00225$, by 1,275 for $\beta = 0.00228$, by 1,368 for $\beta = 0.00231$ and 1,527 if $\beta = 0.00236$. According to Table 6.3, the maximal number of hospitalizations at the turn of the year was reached in the calendar week 51 in 2020 with 11,564 hospitalizations. The height of the peak in the lower right diagram in Figure 6.7 fits this statement of Table 6.3, although the calendar week of the peak is shifted by 2 weeks.

It should be added that the peak is shifted to calendar week 52, but is smaller by around 2,000 individuals in each graph if the parameter z_1 is set to 39 instead of 40. Thus β would have to be increased in order to obtain larger maximal numbers of hospitalized individuals again. If z_1 is set to the value 36, a local maximum of around one third of the respective maximum visible in Figure 6.7 is attained in the calendar week 51, and a maximum of 13,000 to 14,400 per graph is reached in the calendar week 24 in 2021.

Variation of ι

To create Figures 6.8 and 6.9, the lower bound of the parameter β was fixed to the value that yielded the most exact extremum in Figure 6.3 or 6.4, respectively, but the lower bound of the parameter ι was changed within a certain small interval, while q_1 and c_1 were fixed to a value per country each. The purpose has been to find out which value resulted in the most realistic number of individuals in the ICU compartment for the second wave in Sweden or Germany, respectively.

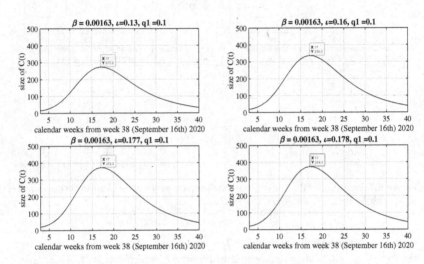

Figure 6.8 Prediction of the size of the ICU compartment for Sweden in the second wave, with $K_2 = 0.061$, $z_1 = 40$, $\iota \in [0.13, 0.178]$

Figure 6.8 conveys that the maximal number of people newly admitted to ICU was reached in the calendar week 4 in 2021, which is 3 weeks delayed compared to Table 6.3, that shows the Swedish ICU admittance peak in the calendar week 1. For the assignment $\iota = 0.13$ the peak is 273 ICU patients, which is depicted in the upper left graph. To give further examples, it is 23% larger for $\iota = 0.16$, 46% larger for $\iota = 0.19$ and 62% larger for $\iota = 0.21$. According to Table 6.3 the most realistic allocation is $\iota = 0.177$ as this leads to the outcome $C(17) = 372$. Finally, a value of slightly less than 50 is achieved in the two upper graphs, and marginally more than 50 is reached in the two lower graphs, in the calendar week 27 in summer.

Figure 6.9 gives an example of how the variation of the contact rate parameter z_1 can influence the size of the ICU compartment. It demonstrates that modifying the parameter z_1 from 40 to 35, the maximal number of Swedish COVID-19 ICU

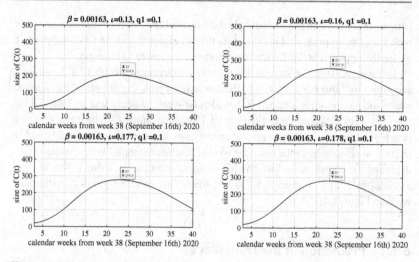

Figure 6.9 Prediction of the size of the ICU compartment for Sweden in the second wave, with $K_2 = 0.061$, $z_1 = 35$, $\iota \in [0.13, 0.178]$

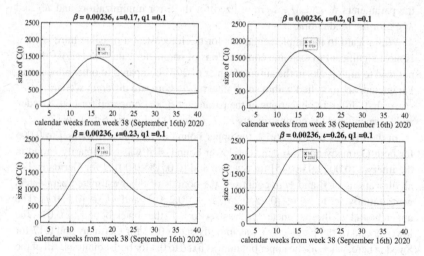

Figure 6.10 Prediction of the size of the ICU compartment for Germany in the second wave, with $K_2 = 0.077$, $z_1 = 40$, $\iota \in [0.17, 0.26]$

patients is achieved in the calendar week 10 instead of 4, all four curves are rounder with a less sharp peak, that is also lower than in Figure 6.8 by 68 for the upper left, 84 for the upper right, 94 for the lower left and the lower right graph.

In Figure 6.10 the most realistic number of German COVID-19 ICU patients according to Table 6.3 is shown in the lower left graph. Nonetheless, the week in which the maximum is attained is the calendar week 52 according to Table 6.3 but calendar week 2 according to Figure 6.10. Finally, a value of slightly less than 500 is achieved in the two upper graphs and marginally more than 500 is attained in the two lower graphs in the calendar week 27 in this example. Whereas the curves depicting the course of the Swedish ICU compartment are declining in the calendar week 27 in 2021 in Figures 6.8 and 6.9, figures are minimally rising in the same week in Figure 6.10.

6.3.2   Prediction of the Third Wave

In the implementations of the first and second wave to predict a possible third wave, the parameters K and M_H were utilized for the error minimization, and K_2 and M_{H2} for the predictions.

With regard to the implementations for the forecast of a possible third wave it is most significant to examine combinations of parameter value allocations that do not lead to high peaks in the autumn/winter of 2021 compared to other allocations. Parameters are modified within small intervals to compare different scenarios.

At first, different allocations of the parameter β are compared for both Sweden and Germany and the compartments I, H, C and D in Figures 6.11 to 6.17.

In the Figures 6.11 and 6.12, the upper bound of the value β was set to 0.005 for both Germany and Sweden. The lower bound of β was exemplarily varied in the interval [0.0022,0.00245] for Sweden and [0.00155,0.0017] for Germany. For all allocations of β and both countries, the bounds of other estimated parameters were initially set to $\varepsilon_H \in [0.05, 0.25]$, $\varepsilon_I \in [0.2, 0.5]$ and $M_\iota \in [0.5, 0.7]$. The upper bound of the parameter q_1 was set to 0.1, the lower bound to 0.01. The lower bound of the parameter ι was initially set to 0.22 for Germany and 0.17 for Sweden here, whereas the upper bound was fixed to 0.4 for both countries. All of the mentioned bounds were selected as they seemed realistic and also yielded reasonable and comparable progressions of future compartment sizes. The parameter c_1 was exemplarily fixed to the value 90 for Sweden and 70 for Germany here. The values initially used for K, M_H and M_{H2} are given in Table 6.1. The parameter K_2 was changed from 0.067 to 0.077 for Germany since it yielded better results for the course of $H(t)$ in the predictions of the first wave.

Variation of β

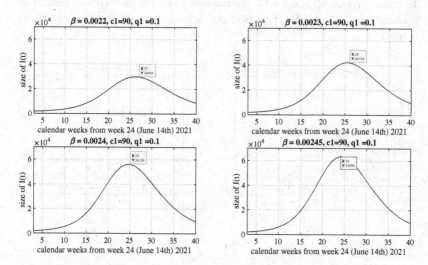

Figure 6.11 Prediction of the size of the infected compartment for Sweden in the third wave, with $\beta \in [0.0022, 0.00245]$

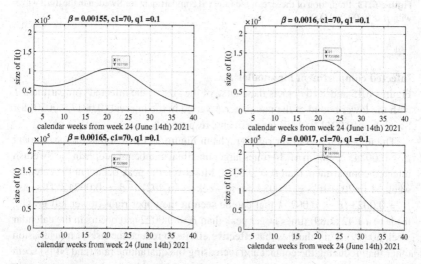

Figure 6.12 Prediction of the size of the infected compartment for Germany in the third wave, $\beta \in [0.00155, 0.0017]$

Infected compartment for larger β

To create Figures 6.13 and 6.14, the transmission rate was increased instead of decreased to show possible impacts of growing numbers of infections with mutations.

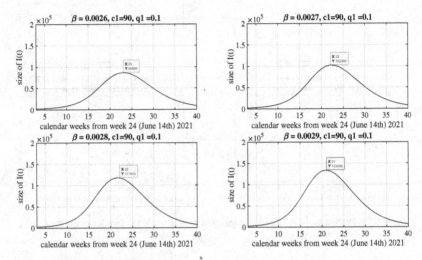

Figure 6.13 Prediction of the size of the infected compartment for Sweden in the third wave, with $\beta \in [0.0026, 0.0029]$

Infected compartment for smaller β

Figures 6.15 and 6.16 depicts the course of the size of the infected compartment I with the lower bound of the parameter β varied in the interval $[0.0017, 0.0020]$ for Sweden or $[0.0011, 0.0017]$ for Germany, respectively.

The upper left and lower right graphs in Figure 6.11 reveal that the assignment $\beta = 0.00245$ results in a 2.14 times larger maximal number of people in the Swedish infected compartment than $\beta = 0.0022$. Moreover, the peak occurs in the calendar week 51 in 2021 if $\beta = 0.0022$ but in week 48 in 2021 if $\beta = 0.00245$. The case $\beta = 0.0023$ ($\beta = 0.0024$), that can be seen in the upper right (lower left graph), leads to a 1.42 (1.89) times larger peak than $\beta = 0.0022$ and occurs in the calendar week 50 (49). This clarifies that effective efforts to reduce the risk of transmission other than reducing the contact and increasing the quarantine rate (and NPIs) seem expedient, since the peak is flattened as well as delayed in the above scenario.

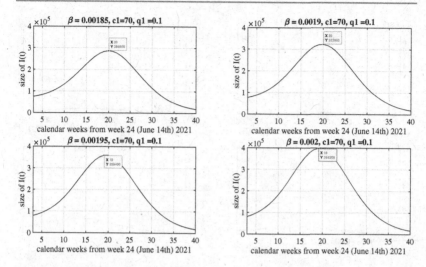

Figure 6.14 Prediction of the size of the infected compartment for Germany in the third wave, with $\beta \in [0.00185, 0.0020]$

Figure 6.12 reveals that an increase in β by $5 \cdot 10^{-5}$ accounts for an increase in the size of the respective reached local maximum in Germany, but not in an earlier occurrence of the respective peak in this parameter scenario. If β is set to 0.0018 or 0.00195 as in Figure 6.14, the maximum is however shifted to the calendar week 44 or 43, respectively. Compared to Figure 6.11 and with respect to the assumed parameter allocations, it is conspicuous that the number of infected people in Germany is predicted to reach its peak 3–6 weeks earlier than Sweden, and in mid-November rather than at the turn of the year 2021/2022.

In Figure 6.13, the progression of the size of the Swedish infected compartment for larger β is presented. It is striking that an increase in β by 10^{-4} consistently leads to a forward shift of the achieved maximum by 1 week as in Figure 6.11. While an assignment of β to a value between 0.00245 and 0.00255 results in a peak of a similar height as the peak reached in the second wave, the allocations in Figure 6.13 account for peak sizes between 86,660 and 133,600. What can be noticed here is that the maximum attained in the lower right graph of Figure 6.13 occurs in the calendar week 45 if $\beta = 0.0029$. This is the same week in which the German maxima are reached in Figure 6.12. With $\beta = 0.0029$, the Swedish maximal size of the infected compartment also has a similar height as the German peak depicted in the upper right graph of Figure 6.12. It has to be considered that a higher contact rate

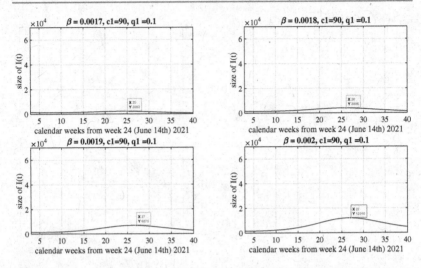

Figure 6.15 Prediction of the size of the infected compartment for Sweden in the third wave, with $\beta \in [0.0017, 0.0020]$

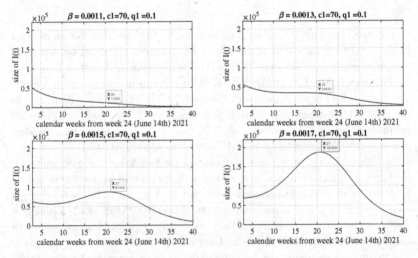

Figure 6.16 Prediction of the size of the infected compartment for Germany in the third wave, with $\beta \in [0.0011, 0.0017]$

compared to Germany, combined with a 0.001 higher transmission risk, is necessary for Sweden to attain this peak height. With the given parameters, the application of the transmission risk $\beta = 0.0029$ leads to a 4.43 times larger and 6 weeks earlier peak in the third wave than $\beta = 0.0022$ for Sweden.

Figure 6.14 demonstrates the same scenario with other assignments of β for Germany. It can be noted that the application of transmission risks larger than about 0.00185 account for the attainments of maxima larger than the maxima reached in the second wave. The peak visible in the upper right graph ($\beta = 0.0020$) is 12% larger than the one in the upper left graph of Figure 6.14 ($\beta = 0.00185$) and 300% larger than the one of the upper left graph of Figure 6.12 ($\beta = 0.00155$).

Figures 6.15 and 6.16 illustrate great contrasts to the Figures 6.13 and 6.14 because much smaller transmission risks are assumed here. The applied transmission rates also yield smaller values than in Figures 6.11 and 6.12. It can be observed that the size of the Swedish infected compartment progresses almost flatly and is on a level of around 2,000 per week over the course of the regarded 40 weeks if $\beta = 0.0017$. The larger β, the clearer a peak becomes. With an assumed $\beta = 0.0020$ the maximal value is already 6 times larger than with $\beta = 0.0017$ and almost 3 times larger than with $\beta = 0.0019$, but 2.5 times smaller than with $\beta = 0.0022$. With regard to Figure 6.15 the allocation $\beta = 0.0011$ results in a consistently decreasing size of the German infected compartment, whereas a slight peak is observable if $\beta = 0.0013$. With $\beta = 0.0015$ the reached maximum is already on a a marginally higher level than the maximum of Sweden in the second wave.

As opposed to the prediction of the course of the size of the infected compartment in the second wave visible in Figure 6.4, all graphs in Figures 6.12, 6.14 and 6.16 are still declining 40 weeks after the start of observation, which is in the calendar week 12 (mid-March) in 2022 here.

It is significant to consider that scenarios of much larger global maxima in the third wave of the pandemic in Europe compared to the second wave are not improbable owing to a growing number of viral mutations leading to enhanced transmission risks. Possible scenarios resemble the ones depicted in Figures 6.13 and 6.14. Certainly, more local peaks, that are not depicted in all of the above figures, are realistic.

Conversely, the general hope with regard to accessible vaccines is to contain transmission rates and as a result the spread of the virus, and flatten worldwide curves as depicted in the upper graphs of Figure 6.16.

The following diagram displays the progressions of the Swedish deceased compartment for the same allocations of β as in Figure 6.11. It has to be considered that a modification of the sizes of the parameters M_H and M_ι would certainly account for changes in the size of the deceased compartment, but publicly available values for these parameters were applied here. In Figure 6.17, the influence of β on the number of deceased COVID-19 cases is investigated.

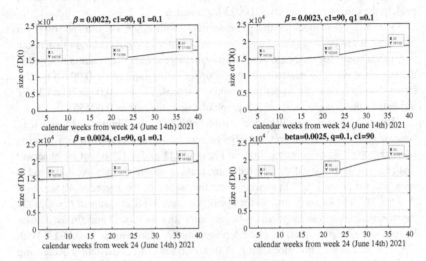

Figure 6.17 Prediction of the size of the deceased compartment for Sweden in the third wave, with $\beta \in [0.0022, 0.00245]$, $K_2 = 0.061$, $M_t = 0.5$

According to the used data set, approximately 12,500 people were confirmed COVID-19 deaths in Sweden until the calendar week 6 in 2021. A CFR of 0.0432 is applied to obtain the assumption that there will be 14,730 confirmed Swedish COVID-19 deaths by calendar week 24 in 2021. Figure 6.17 demonstrates the different growths of the size of the deceased compartment if different transmission risks β are applied. Whereas the total number of confirmed COVID-19 deaths is almost equal in the calendar week 44 (first week of November) in all four graphs of the figure, great differences are observable in the calendar week 7 in 2022. The high hospital occupancy rate and probable hospital and ICU overburdening, that is reflected by high mortality rates in certain weeks in the underlying data set, is re-recognized in the comparatively strong increases after calendar week 44 in Figure 6.17.

The assignment of β to the value 0.0022 leads to a total of 17,180 COVID-19 deaths in Sweden in the implementation. An enlargement of β by 0.001, 0.002 or 0.003 from the value 0.0022 results in an increase in the reached maximal value of 5.4%, 11.6 or % 18%, respectively.

Variation of ε_H and ε_I, Germany

For the creation of Figure 6.18, the lower bound of the parameter β was fixed to the value 0.00155, but the parameters ε_I and ε_H were varied within certain small intervals. As regards content, the probability of a transmission emerging from an individual in the compartment I or H compared to the probability of a transmission emerging from an infectious undetected case was modified. Here, the bounds of the parameters ι, M_ι and q_1 were set to their mentioned initial values.

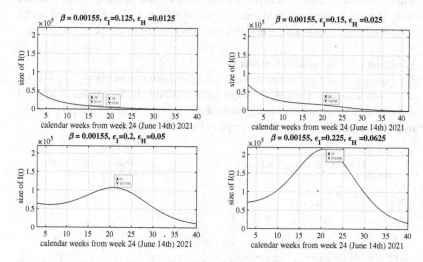

Figure 6.18 Prediction of the size of the infected compartment for Germany in the third wave, with $\varepsilon_H \in [0.0125, 0.0625]$, $\varepsilon_I \in [0.125, 0.225]$

With reference to the two upper curves in Figure 6.18 it becomes clear that the combination of the assignments $\varepsilon_I = 0.125, \varepsilon_H = 0.0125$ or $\varepsilon_I = 0.15, \varepsilon_H = 0.025$ results in declining graphs. Less than 5,000 infected people in the compartment I are achieved around the calendar week 43 in 2021 in the upper left and the calendar week 1 in 2022 in the upper right graph.

The lower left graph in Figure 6.18 is equal to the upper left one in Figure 6.12. Increasing ε_I from 0.15 to 0.2 and ε_H from 0.025 to 0.05 in the lower left graph yields a curve that increases from the calendar week 28 until 45, reaches a peak then (beginning of November), and finally decreases stronger than it increased before. Therefore, with the assumed parameters, the question whether the transmission risks emerging from the compartment I (and H)is 15 or 20% (2.5 or 5%) of the transmission risk emerging from undetected infectious people is significant for the progression of the size of the infected compartment. This finding is already deducible from the sensitivity analysis in Table 5.6.

What is most conspicuous in the two lower graphs of Figure 6.18 is the fact that another increase in ε_I by 12.5% and ε_H by 25% leads to a duplication of the maximally reached size of the infected compartment in the calendar week 21. This stresses the high relevance of an effective isolation in the case of infectedness in order to more drastically reduce the risk of transmission originating from confirmed infected cases in comparison to unconfirmed ones. In this scenario, the authorities and population should make every effort to reduce case-fatality rates.

Variation of β, ε_H and ε_I, Germany

For Figure 6.19, the parameter β was varied per value combination of ε_I and ε_H in the way that approximately the same local maximum in the number of infected individuals was attained per assignment of β, ε_H, ε_I. The lower left graph in the figure is the same as the upper left graph in Figure 6.12 again.

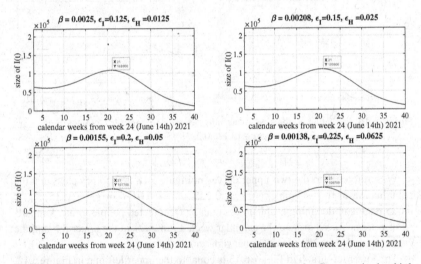

Figure 6.19 Prediction of the size of the infected compartment for Germany in the third wave, with $\beta \in [0.00138, 0.0025]$, $\varepsilon_H \in [0.0125, 0.0625]$, $\varepsilon_I \in [0.125, 0.225]$

In all four scenarios, a peak between 107,700 and 109,800 is attained in the calendar week 45. It is illustrated that the size of β has to be decreased by 16.8% if $\varepsilon_I = 0.15$, $\varepsilon_H = 0.025$ is applied instead of $\varepsilon_I = 0.125$, $\varepsilon_H = 0.0125$ to attain approximately the same peak size. Similarly, β has to be decreased by 25.5% if $\varepsilon_I = 0.2$, $\varepsilon_H = 0.05$ is applied instead of $\varepsilon_I = 0.15$, $\varepsilon_H = 0.025$, and by 11% if $\varepsilon_I = 0.225$, $\varepsilon_H = 0.0625$ is applied instead of $\varepsilon_I = 0.2$, $\varepsilon_H = 0.05$ to attain the same target.

With respect to content, this means that an increased general transmission risk, which is especially caused by spreading mutations, has to be compensated by provisions like enhanced vaccination programs, which in the case of administered leaky vaccines in the $SIHCDR$ model results in a decrease in the transmission risk emerging from all infected compartments. This was indicated in Subsection 4.3.3. The corresponding target would be to remain on the same level concerning the predicted size of the infected class.

As opposed to leaky vaccines, an available all-or-nothing vaccine would result in the exclusion of vaccinated individuals from the susceptible state. Thus the susceptible population would simply become smaller in the $SIHCDR$ model. A better testing strategy would effect a lower dark figure as well as a larger quarantine rate.

In the sequel, the behaviour of the compartment size curves with regard to changes in the assignments of the parameters q_1 and c_1 is analysed. NPIs, which are implemented and varied by the authorities, are included in the modification of the parameters apart from sole quarantine measures.

A scenario which is characterized by no widely implemented quarantine program but mainly the increased awareness in response to the pandemic itself and initial recommendations given by the media or health institutions does not result in a considerable change in the parameter q_1 but in a reduction of c_1. This can be classified as a scenario of **minimal intervention** [78]. Self-quarantine is possible here. Nonetheless, the allocation $q_1 = 0$ stands for a scenario in which only a negligible fraction of the population is in quarantine.

As opposed to this, the adoption of control measures as the closure of schools, universities, restaurants and other facilities can be regarded as putting the people affected by this in a "partial quarantine". This kind of quarantine is not imposed directly on the affected people by the state or an institution and is no self-quarantine, but ensures that the individuals have less contacts and leave the house less often. Throughout the following analyses of generated predictions, this is described as the **baseline scenario** and assumed to equal the allocation of q_1 to the value 0.1 and strongly reduced contact rates compared to a scenarios with an absent pandemic. Assigning q_1 to a value in the interval $(0, 0.1)$ can be regarded as a scenario of light NPIs.

A scenario of an economic shut-down can be classified as a scenario of a **lock-down**. It incorporates high vigilance in the form of an extended testing activity leading to more case isolations as well as more imposed quarantines to contact persons of infected cases [78]. Moreover, curfews, strong contact restrictions, remote working in all sectors in which it is possible are realized. The aim to contain the pandemic is realized as strongly as possible in a regarded country. This scenario is implemented here by assigning q_1 to a value in the interval $(0.1, 0.5)$. Depending on the stringency of the curfews and restrictions, for instance the time span during which individuals are not allowed to go out or the number of people a person is still allowed to meet, the parameter q_1 can be adjusted.

It is assumed that $q_1 = 0.5$ symbolizes a scenario that is even more effective than a lockdown in terms of protecting individuals from contagion. For instance, it can be imagined that the dark figure is diminished since a larger part of infections is detected through testing, such that family members and other contact persons are put in quarantine more often. An extended vaccination program including accessible vaccines for a large and growing share of the population rather leads to a reduced transmission rate than a reduction of q_1. It might indirectly influence the quarantine rate since quarantine regulations may be adapted to the general vaccination status of the population. A temporary relaxation of the applied intervention measures is not assumed in this scenario. A variation of the contact rate parameters c_1 and c_2 usually accompanies the implementation of quarantine measures, since interventions initialized by authorities normally affect both contact and quarantine rates in the population. In the following analysed scenarios, the quarantine and contact rate are firstly modified independently of each other in order to examine the effect that the variation of each of the rates has.

Variation of q_1
Primarily, the parameter q_1 was modified in the interval $[0,0.5]$ in order to display scenarios of different quarantine measures, that could include a lockdown or other interventions of a certain stringency. The results for the course of the sizes of the compartments I, H and C are depicted in Figures 6.20 to 6.24. In the analysis of Figures 6.20 and 6.21 it has to be considered that the contact rate parameter c_1 is fixed to a value, which is already slightly lower than in a scenario of an absent COVID-19 pandemic. Subsequently, the graph resulting from setting $q_1 = 0$ does not refer to a scenario without intervention, but without a realized quarantine program. The parameter β has the value 0.00235 for Sweden in Figure 6.20 and 0.00165 for Germany in Figure 6.21.

Figures 6.20 and 6.21 convey that the missing application of quarantine measures effects a comparatively very high number of infected individuals. In both figures,

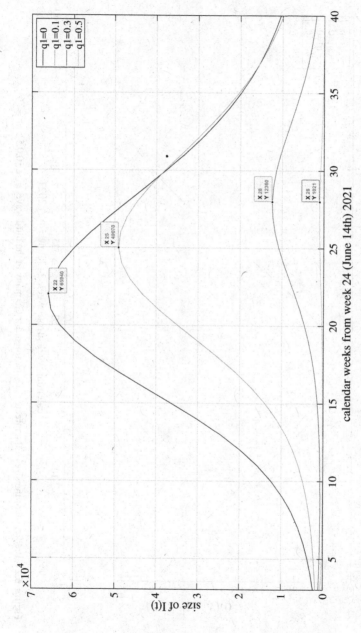

Figure 6.20 Prediction of the size of the infected compartment for Sweden in the third wave, with $q_1 \in [0, 0.5]$, $c_1 = 90$

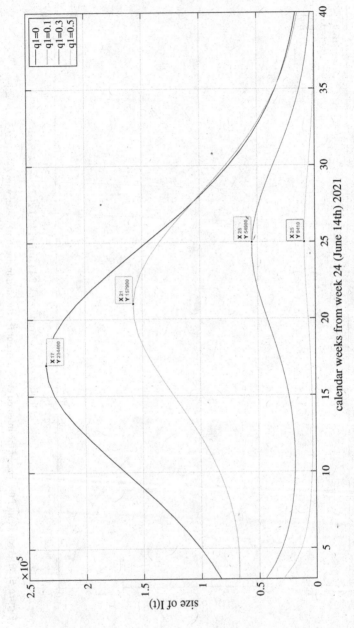

Figure 6.21 Prediction of the size of the infected compartment for Germany in the third wave, $q_1 \in [0,0.5]$, $c_1 = 70$

the maximal number of infected individuals is reached earlier than with applied quarantine regulations.

In the scenario shown in Figure 6.20, the peak is larger by about 16,000 infected individuals and attained 3 weeks earlier if no Swedish inhabitant is put in quarantine than in the baseline scenario. The scenario of $q_1 = 0.1$ may include precautious self-quarantine. An extreme drop of around 38,000 infected individuals at the peak as well as a delay of the peak by 3 weeks is realized if the parameter q_1 is set to the value 0.3 instead of 0.1. Here, an at least partial lockdown is assumed instead of the baseline scenario.

Whereas 234,400 individuals are in the infected compartment in the calendar week 51 in 2021 if q_1 is set to zero, a peak of 157,900 (33% less) is attained in the calendar week 3 in 2022 in the baseline scenario. As in the Swedish scenario, the height of the peak is extremely diminished if a lockdown is assumed ($q_1 = 0.3$ here). The maximal size of the infected compartment is reduced by a third and the peak is shifted by 4 weeks compared with the baseline scenario.

If the parameter q_1 is changed to the value 0.5, no clear peak is discernible in both figures. The resulting graphs are almost flat. These scenarios are certainly very unlikely to reach, since lockdown measures that are not eased over more than half a year are improbable as they are difficult to implement and almost incompatible with human needs. Nonetheless, a permanent availability of inexpensive tests and constant testing of people who live in, or visit high risk regions or were or are going to be in a closed room with people other than their housemates for more than a defined time period (for instance 15 minutes) favours scenarios with very low infection numbers.

The analysis implies that a lockdown and enhanced testing facilities are generally justified and favourable measures to reduce transmissions. The interventions included in the lockdown do not have to be exceedingly strict to achieve an important decline in the weekwise numbers of infections according to the above two figures. It is realistic that they are adapted to the current incidences monthwise, weekwise and sometime daywise and temporarily differ between the (federal) regions of a country. An optimal quarantine program algorithm could precisely estimate which intervention yields the perfect balance between the protection of the population and containment of the viral spread and the fulfilment of the human need for social contacts per region, government area or city without any time delay.

Similar observations as for the course of the infected compartment can be made for the hospitalized and ICU compartments. The progression of the ICU compartment is depicted for Sweden in Figure 6.22 and Germany in Figure 6.23 for the same parameter allocations as in Figures 6.20 and 6.21.

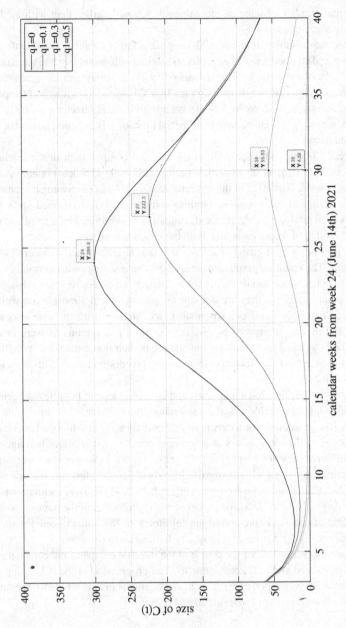

Figure 6.22 Prediction of the size of the ICU compartment for Sweden in the third wave, with $\beta = 0.00165$, $\varepsilon_I = 0.2$, $\varepsilon_H = 0.05$, $\iota = 0.2$, $q_1 \in [0,0.5]$, $c_1 = 90$

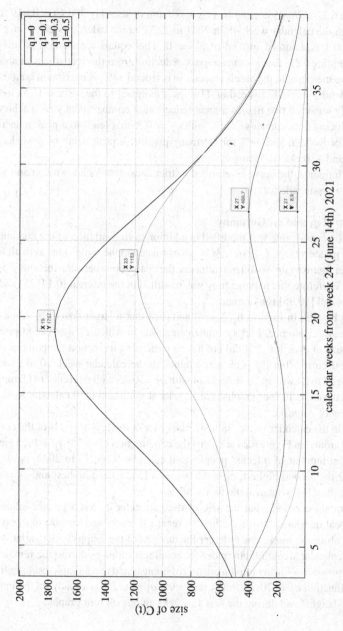

Figure 6.23 Prediction of the size of the ICU compartment for Germany in the third wave, with $\beta = 0.00165$, $\varepsilon_I = 0.2$, $\varepsilon_H = 0.05$, $\iota = 0.2$, $q_1 \in [0, 0.5]$, $c_1 = 70$

Figure 6.22 shows a decline from a maximum of scarcely 300 COVID-19 ICU patients in the calendar week 48 in 2021 to 222.2 in the calendar week 1 in 2022 if $q_1 = 0.1$ is assumed instead of $q_1 = 0$. This equals a reduction of around 26%. In Figure 6.23, the percentage peak reduction from the scenario with absent quarantine measures to the baseline scenario is around 33%. A maximum is attained in the calendar week 47 instead of 43 if q_1 is assigned to the value 0.1. As in the previously regarded two figures a great reduction in compartment size is achieved by introducing a lockdown scenario with $q_1 = 0.3$, that leads to a peak reduction of 25% for Sweden and 34% for Germany just like a peak delay of 3 weeks for Sweden and 4 weeks for Germany.

In both figures, the curve levels off if a strict consistent lockdown scenario with $q_1 = 0.5$ is assumed.

Variation of q_1 and c_1, Germany

Finally, the contact rate was modified in addition to the variation of the quarantine rate. The parameters β, ε_I, ε_H, ι, M_t were estimated to the same value as in all four graphs per figure again, so did not influence the distinctions between the four graphs per plot. Therefore, the parameter q_1 was modified in the interval [0.1,0.25] and c_1 in the interval [80,95] for Germany.

It can be seen in Figures 6.22, 6.23 and 6.24 that a larger size of q_1 results in a later peak occurrence. Further implementations in which c_1 was varied proved that the size of $c_1 \in [60, 75]$ did not have a noticeable impact on the point in time the peak occurred, but the peak was attained in the calendar week 20 to 21 with $q_1 = 0.1$, 23 to 24 with $q_1 = 0.2$ and 24 with $q_1 = 0.25$ with reference to Germany.

In Figure 6.24, all four graphs achieve a local minimum until calendar week 32 (beginning of August).

Only in the calendar weeks 38 to 42, the order of the height levels of the curves is turned around in Figure 6.24 such that the combination $c_1 = 75, q_1 = 0.25$ yields the highest number of infected people until calendar week 12 in 2022, $c_1 = 70$, $q_1 = 0.2$ the second highest, $c_1 = 65, q_1 = 0.15$ the third highest and $c_1 = 60$, $q_1 = 0.1$ (baseline scenario) the fourth highest.

One might conclude that the size of the parameter q_1 has a greater influence on the local minimum during the first 15 regarded weeks and the size of c_1 on the other 25 observed weeks due to larger fluctuations of the graphs under variation of q_1 until calendar week 39, but under the modification of c_1 during the remaining weeks. However, a further implementation demonstrated that a graph resulting from the combination $c_1 = 70, q_1 = 0.1$ ($c_1 = 95, q_1 = 0.25$) is already on the lowest (highest) height level during the first 15 weeks among all four graphs.

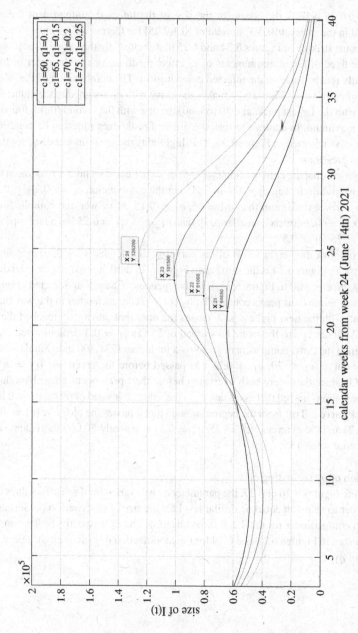

Figure 6.24 Prediction of the size of the infected compartment for Germany in the third wave, with $\beta = 0.00165$, $\varepsilon_I = 0.2$, $\varepsilon_H = 0.05$, $\iota = 0.2$, $q_1 \in [0.1, 0.25]$, $c_1 \in [60,75]$

Figure 6.25 depicts the course of the size of the infected compartment for q_1 modified in the interval $[0,0.3]$ instead of $[0.1,0.25]$ for Germany.

It is most striking in Figures 6.24 and 6.25 that exclusively the varied allocations of q_1 for fixed different assignments of c_1 effect modifications in the order of the maximally reached size of the infected compartment. The combination of the allocations of c_1 and q_1 in Figure 6.24 makes the curve with c_1 assigned to the value 75 the one with the largest peak, and 70 (65, 60) the one with the second (third, fourth) highest maximum. In contrast to this, the order of peak sizes sorted in descending order of c_1 is reversed in Figure 6.25. This highlights the great impact of q_1 on the graph progressions.

It is visible that the combination $c_1 = 65, q_1 = 0.1$ causes a higher maximum in the given conditions than $c_1 = 60, q_1 = 0$, but the combination $c_1 = 60, q_1 = 0.1$ leads to a lower maximum than $c_1 = 65, q_1 = 0.15$. Moreover, the combination $c_1 = 70, q_1 = 0.2$ effects a smaller peak than $c_1 = 75, q_1 = 0.25$ but a larger peak than $c_1 = 75, q_1 = 0.3$.

The order of the height levels of the curves until calendar week 32 (beginning of August) in Figure 6.24 is the same as in Figure 6.25 until approximately calendar week 44. With regard to Figure 6.25, it is conspicuous that the four depicted graphs do not only attain their peaks consecutively, but overtake each other in the way that the graph with the next highest peak passes the ones that previously reached their peaks. Assigning c_1 to the value 85 instead of 75 for $q_1 = 0.3$ demonstrated that the attained maximal compartment size was a lot higher (231,300 individuals), and the blue curve ($c_1 = 70, q_1 = 0.2$) was passed before the green one ($c_1 = 65, q_1 = 0.1$) since they were both overtaken before their peak occurrences. Another implementation showed that assigning c_1 to the value 50 instead of 60 for $q_1 = 0$ let the black curve fall off from the beginning such that it passes the blue one ($c_1 = 70, q_1 = 0.2$) and the cyan one ($c_1 = 75, q_1 = 0.3$) at scarcely 50,000 individuals in the calendar week 13.

Variation of q_1 regarding c_1

Regarding Figures 6.26 or 6.27, the parameter q_1 was varied for 4 selected values of c_1 in order to attain an equal or similar level for the size of the Swedish or German infected compartment for the 4 combinations of q_1 and c_1, respectively. The same was realized in Figures 6.28 and 6.29 for the modification of c_1 for certain selected values of q_1.

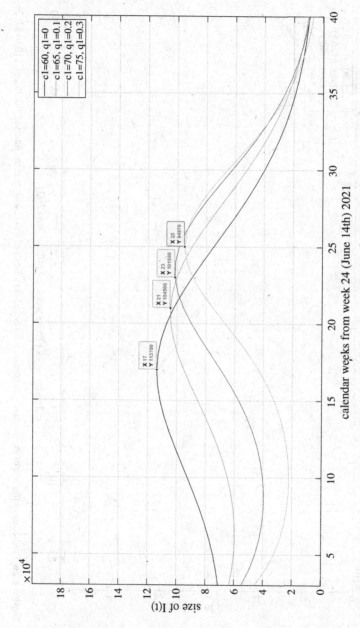

Figure 6.25 Prediction of the size of the infected compartment for Germany in the third wave, with $\beta = 0.00165$, $\varepsilon_I = 0.2$, $\varepsilon_H =$ 0.05, $\iota = 0.2$, $q_1 \in [0,0.3]$, $c_1 \in [60,75]$

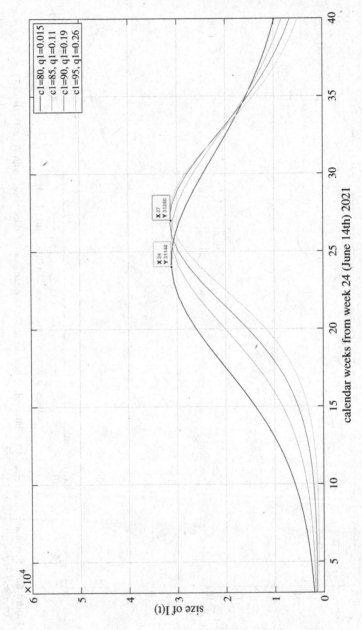

Figure 6.26 Prediction of the size of the infected compartment for Sweden in the third wave, with c_1 in the interval [80,95], $q_1 \in [0.015,0.26]$

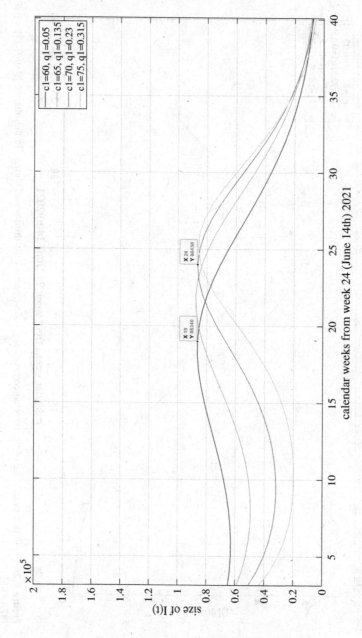

Figure 6.27 Prediction of the size of the infected compartment for Germany in the third wave, with c_1 in the interval $[60, 75]$, $q_1 \in [0.05, 0.315]$

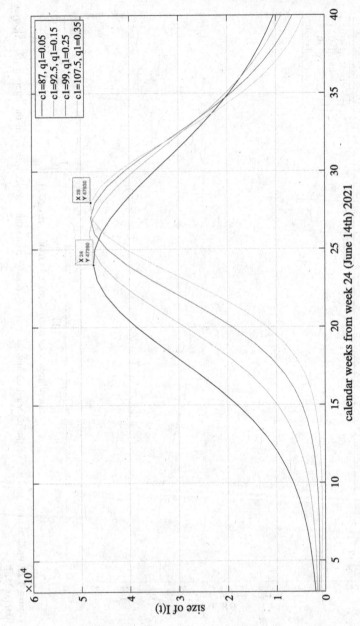

Figure 6.28 Prediction of the size of the infected compartment for Sweden in the third wave, with q_1 in the interval [0.05,0.35], $c_1 \in$ [87,107.5]

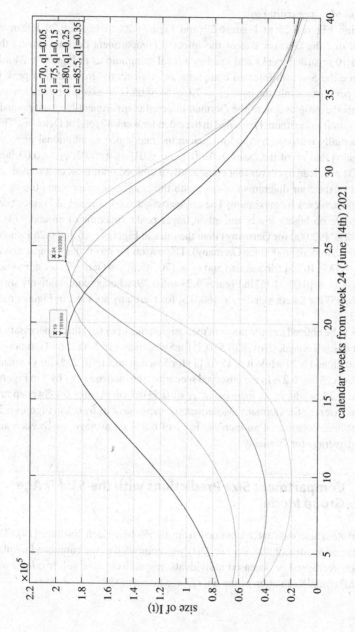

Figure 6.29 Prediction of the size of the infected compartment for Germany in the third wave, q_1 in the interval $[0.05, 0.35]$, $c_1 \in [70, 85.5]$

Variation of c_1 regarding q_1

Comparing Figure 6.26 to Figure 6.27 and Figure 6.28 to Figure 6.29 it can be observed that the German size of the infected compartment declines during the first 5 to 10 regarded weeks and reaches a local minimum as in Figures 6.24 and 6.25 while the Swedish infected compartment consistently grows until a peak is attained per curve. Only the case $c_1 = 70, q_1 = 0.05$ is an exception here since it represents the progression of the German infected compartment but continuously increases until a maximum is reached in the calendar week 43 (end of October). This is substantially reasoned by the low quarantine rate, since an additional program run revealed that all of the cases $q_1 = 0, q_1 = 0.01, q_1 = 0.02, q_1 = 0.03$, and $q_1 = 0.04$ also lead to curves that consistently increased until peak occurrence.

In all of the four diagrams the case with the smallest q_1 represents the curve that firstly achieves its maximum, The compartment sizes reached in Figures 6.28 and 6.29 are on higher levels and attain larger peaks (maxima of around 47,000 for Sweden, 192,000 for Germany) than the ones in Figures 6.26 and 6.28 (around 31,000 for Sweden, 86,000 for Germany). The reason for this is that comparatively large $c_1 \in [87, 107.5]$ for Sweden and $c_1 \in [70, 85.5]$ for Germany are combined with small $q_1 \in [0.05, 0.35]$ in Figures 6.28 and 6.29, whereas comparatively small $c_1 \in [80, 95]$ for Sweden and $c_1 \in [60, 75]$ for Germany are used in Figures 6.26 and 6.27.

It has to be noticed that the range in the assignments of c_1 (q_1) that is necessary to attain very similar peak sizes is 20.5 (0.2) for Sweden and 15.5 (0.2) for Germany in Figures 6.28 and 6.29 while it is 15 (0.11) for Sweden and 15 (0.265) for Germany in Figures 6.26 and 6.27. In the observed scenarios, the increase of c_1 by 5 in Figures 6.26 and 6.27 results in an increase in q_1 of 0.53167 on average for Sweden and 0.1033 on average for Germany. Moreover, the increase of q_1 by 0.1 in Figures 6.28 and 6.29 is responsible of an increase in c_1 of 6.833 on average for Sweden and 5.167 on average for Germany.

6.4   Compartment Size Predictions with the $SVID$ Age Group Model

Compartment size data were obtained from the Robert-Koch Institute [14]. The weekwise compartment size $V(t)$ or $D(t)$ was computed as the cumulated number of weekly recovered or deceased individuals, respectively. The solver $ode45$ was used for the implementation of the $SVID$ age group model.

In each program run, three different allocations of the parameter v_b were selected per age group, which were saved in the vectors v_{b1}, v_{b2}, v_{b3}. With regard to the figures below, it is defined that an entry $v_{bi}(1)$ is used for the age group of 0–19-year-old people, $v_{bi}(2)$ for the age group of 20–59-year-old individuals and $v_{bi}(3)$ for the age group of over 60 year-old people for all $i \in \{1, 2, 3\}$.

It has to be taken into consideration that absolute numbers of individuals per compartment are depicted as in the previous diagrams, but the three regarded age groups have different sizes in terms of associated individuals. For a better evaluation of the relative share of infected, vaccinated and deceased individuals per age group and week in the following diagrams, it can be assumed that the population group of 0–19-year-old people has a size of 13,500,000, the group of 20–59-year-old people has a size of 35,900,000 and the group of over 60-year-old people has a size of 19,800,000 individuals. Those numbers are assumed to approximately equal the sizes of the corresponding German population groups containing those who want to get vaccinated.

Figures 6.30 to 6.32 display the progressions of the sizes of the vaccinated, infected or deceased compartment, respectively, per age group and for each of the parameter vectors $v_{b1} = [13, 28, 33]$, $v_{b2} = [11, 25, 30]$, $v_{b3} = [8, 18, 24]$. The quarantine rate q was estimated as 0.2 for the 0–19-year-old people, 0.3 for the 20–59-year-old people, 0.1 for the over 60 year-old people in this scenario. The modification factor ε_I was estimated as 0.15 for all three age groups. Furthermore, the transmission rate β was estimated as 0.025. The quarantine ratio was estimated as 0.2 for the 0–19-year-old people, 0.3 for the 20–59-year-old people, 0.1 for the over 60 year-old people in this scenario. The modification factor ε_I was estimated as 0.15 for all three age groups.

According to Figure 6.30 and observing the three graphs created with v_{b3}, all individuals who want to be vaccinated in the age group of over 60-year-old people will have received their second vaccination by calendar week 4 in 2022, and all individuals in the age group of 20–59 (0–19-)-year-old people by calendar week 34 (20) in 2022. Hence in this scenario, the over 60-year old age group is almost entirely vaccinated one year and the 20–59-year-old group more than one and a half year after a second COVID-19 vaccination was initially given in Germany. If the parameter vector v_{b2} or v_{b1} is selected instead of v_{b3}, only an exceedingly small, almost not visible flattening of the correspondent curve of the elderly age group can be noticed. The graph corresponding to v_{b2} (v_{b1}) of the 0–19-year old group saturates around the calendar week 36 (45) in 2022, whereas the graph corresponding to v_{b2} or v_{b1} of the 20–59-year old group attains a maximal level after the first calendar week in 2023 in this case.

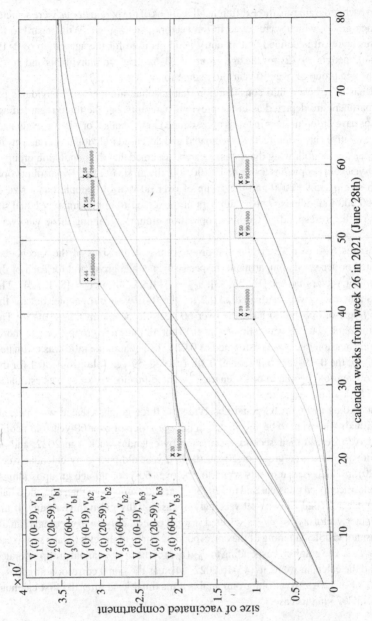

Figure 6.30 Prediction of the size of the vaccinated compartments of three age groups in the third wave with $\beta = 0.025$, for each of the parameter vectors $v_{b1} = [13, 28, 33]$, $v_{b2} = [11, 25, 30]$, $v_{b3} = [8, 18, 24]$

Figure 6.30 shows that more 20–59-year-old than over 60-year-old people will have been vaccinated in total as of calendar week 5 in 2022 if v_{b3} is assumed, calendar week 12 if v_{b2} is chosen and calendar week 14 if v_{b1} is selected.

The following figure illustrates the forecasts created with the mentioned different vaccination rates concerning the size of the infected compartment for all three age groups.

It is striking in Figure 6.31 that a maximum for the 20–59-year-old (0–19-year-old) people is achieved 2.67 (3) weeks on average later than a maximum of the elderly group. The earlier decrease of the 3 visible graphs of the over 60-year-old group is among others accounted for by the earlier achievement of nationwide realized vaccinations.

The figure also conveys that the average modification of the maximally reached number of people in the infected compartment is –7.6% for the 0–19-year-old, –7% for the 20–59-year-old and –5% for the over 60-year-old group if the graphs created with v_{b3} are compared to those established with v_{b1}. This stresses the significance of extensive vaccination programs that pursue the target of providing a protective effect against COVID-19 to a fast growing share of the population.

Additionally, the difference in the calendar week in which the respective peak is attained is –1 between v_{b1} and v_{b2}, –1 between v_{b2} and v_{b3} for the 0-19-year-old group, zero between v_{b1} and v_{b2}, -1 between v_{b2} and v_{b3} for the 20–59-year-old group, and zero between v_{b1} and v_{b2}, zero between v_{b2} and v_{b3} for the over 60-year-old group. This indicates that a small delay of the peak can be achieved apart from an attenuation of the incidence level if more people are vaccinated per unit of time.

It can be computed that the total given number of infected people is 279,610 in the calendar week 51, 283,980 in the calendar week 50, 284,620 in the calendar week 49, 280,830 in the calendar week 48 and 273,460 in the calendar week 47 if the curves resulting from v_{b3} are taken as a basis. Hence, as in most of the predictions for the German infected compartment in Section 6.3, it is predicted here that a maximum is attained shortly before the Christmas week in 2021.

It can be seen in Figure 6.32 that the number of deceased individuals per age group is similar for the 3 allocations of the parameter vector v_b. Concerning the youngest regarded age group, which starts at 20 deaths in the calendar week 26, the compartment size remains on a very low level over the course of observed weeks. The deceased compartment of the 20–59-year-old people comprises 16,340 individuals in the calendar week 29 in 2021 and grows until it contains 17,260 dead people in the calendar week 4 in 2022, which account for a growth rate of 5.63%. The greatest total as well as relative increase in case-fatality is obviously given in the over 60-year-old group since there are scarcely 100,000 deceased ones in the

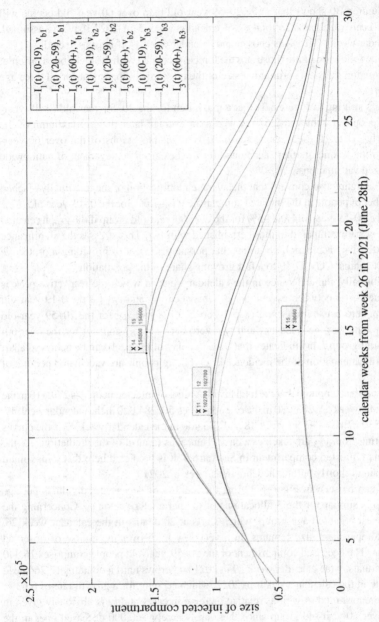

Figure 6.31 Prediction of the size of the infected compartments of three age groups in the third wave with $\beta = 0.025$, for each of the parameter vectors $v_{b1} = [13, 28, 33]$, $v_{b2} = [11, 25, 30]$, $v_{b3} = [8, 18, 24]$

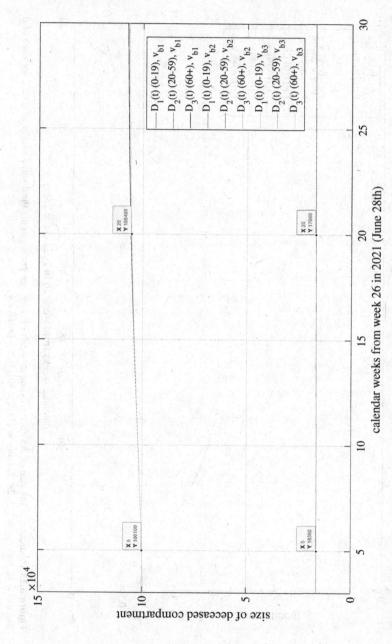

Figure 6.32 Prediction of the size of the deceased compartments of three age groups in the third wave with $\beta = 0.025$, for each of the parameter vectors $v_{b1} = [13, 28, 33]$, $v_{b2} = [11, 25, 30]$, $v_{b3} = [8, 18, 24]$

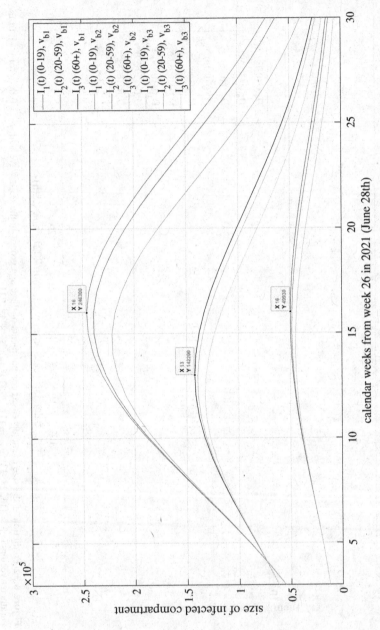

Figure 6.33 Prediction of the size of the infected compartments of three age groups in the third wave with $\beta = 0.027$, for each of the parameter vectors $v_{b1} = [13, 28, 33]$, $v_{b2} = [11, 25, 30]$, $v_{b3} = [8, 18, 24]$

calendar week 29 in 2021, 104,000 in the calendar week 41 in 2021 and 107,400 in the calendar week 4 in 2022. This reveals a growth rate of 7.4% from the calendar week 26 in 2021 until the calendar week 4 in 2022.

Certainly, it has to be taken into consideration that a local maximum in the SARS-CoV-2 incidence is attained within the 30 observed weeks according to Figure 6.31, which leads to a peak in death numbers and is potentially a global maximum with reference to the beginning of the pandemic in February 2020 until the beginning of the year 2022.

In the following scenario, the transmission risk β was set to the value 0.027 instead of 0.025.

Figure 6.33 reveals similar relations between the graphs created with v_{b1}, v_{b2} and v_{b3} as Figure 6.31. However, it is conspicuous that the 9 visible achieved peaks are all larger than the corresponding ones in Figure 6.31, which is reasoned by the larger transmission risk β.

The average absolute difference in peak sizes between the graphs established with $\beta = 0.025$ and $\beta = 0.027$ is 17,980 for the 0–19-year-old, 85,733 for the 20–59-year-old and 37,203 for the over 60-year-old individuals.

The average modification of the maximally reached number of people in the infected compartment is –10% for the 0–19-year- and 20–59-year-old and –7% for the over 60-year-old group if the graphs created with v_{b3} are compared to those established with v_{b1}. Thus larger β lead to bigger differences in the maxima attained with the regarded smaller compared to larger assignment of the parameter v_b. This implies a greater necessity of extensive vaccinations if a greater transmission risk is present.

6.5   Application of Non-Standard Solvers

Nonstandard finite difference (NSFD) scheme methods for the numerical integration of differential equations had their origin in a paper by R. Mickens published in 1989 [277]. They are a generalization of the usual discrete models of differential equations. Furthermore, they preserve certain properties like the positivity or the asymptotic behaviour of the analytic solution of differential equations on the discrete level. Their most important characteristic is, in many cases, the complete absence of the elementary numerical instabilities which plague common finite difference schemes [277].

In Subsection 6.5.1, the NSFD scheme is defined in general and established for the $SVIHCDR$ model. Subsection 6.5.2 shows the results of the implementation of this scheme in $MATLAB$ for the purpose of comparing it with the results obtained in Subsection 6.3.2.

6.5.1   Definition of Nonstandard Finite Difference Schemes

To understand *nonstandard finite difference schemes*, the definition of an *exact finite difference scheme* is necessary. Finite difference methods are a class of numerical techniques to solve differential equations by approximating derivatives with finite differences. Derivatives of differential equations are replaced by difference quotients based on a regular grid in the finite difference method. Nonlinear systems of ODEs are turned into systems of linear equations. If one spatial dimension is considered, a derivative can be approximated by one of the following difference quotients as

$$u'(t) \approx D_h^+ u(t) := \frac{u(t+h) - u(t)}{h} \text{ (first-order forward difference quotient) },$$

$$u'(t) \approx D_h^- u(t) := \frac{u(t) - u(t-h)}{h} \text{ (first-order backward difference quotient) },$$

$$u'(t) \approx D_h^0 u(t) := \frac{u(t+h) - u(t-h)}{2 \cdot h} \text{ (first-order central difference quotient) },$$

where $u(t)$ is the exact solution of the regarded differential equation and h is the step size.

The second-order difference quotients are defined as

$$u''(t) \approx D_h^{2,+} u(t) := \frac{u(t+2 \cdot h) - 2 \cdot u(t+h) + u(t)}{h^2}$$

(second-order forward difference quotient) ,

$$u''(t) \approx D_h^{2,-} u(t) := \frac{u(t) - 2 \cdot u(t-h) + u(t-2 \cdot h)}{h^2}$$

(second-order backward difference quotient),

$$u''(t) \approx D_h^2 u(t) := \frac{u(t+h) - 2 \cdot u(t) + u(t-h)}{h^2}$$

(second-order central difference quotient).

As the accuracy, i.e. the convergence of the approximate solution u of a discrete problem to the exact solution y, of finite difference schemes is a fundamental area of interest, truncation errors of finite difference schemes concerning applied numerical methods are usually calculated. This computation is normally pursued by computing the difference between the regarded finite difference quotient and the Taylor series of $u(t)$ around the point t_n where the derivative is evaluated. A difference scheme is exact if the truncation error is equal to zero or the approximate solution y equals the exact solution u at the grid nodes [278]. Numerical algorithms of high order accuracy can be constructed on the basis of *exact finite difference schemes* [278].

The extension and generalization of detailed studies concerning exact finite difference schemes to special groups of differential equation, which exact schemes are not available for, provide information with respect to the required properties of NSFD methods [277]. A numerical scheme for a system of first-order differential equations is called NSFD scheme if at least one of the following conditions described in [277] is satisfied:

- The first-order derivatives in the system are approximated by the generalized forward difference method (forward Euler method) $\frac{du_n}{dt} \approx \frac{u_{n+1}-u_n}{\phi(h)}$, where $u_n = u(t_n)$ and $\phi \equiv \phi(h)$ is the so-called *denominator function* such that $\phi(h) = h + \mathcal{O}(h^2)$.
- The nonlinear terms are approximated in a non-local way, for instance by a suitable function of several points of a mesh, like $u^2(t_n) \approx u_n u_{n+1}$ or $u^3(t_n) \approx u_n^2 u_{n+1}^2$.

According to Mickens [277], further basic rules of NSFD are the equivalence between the orders of the discrete derivatives and the orders of the corresponding derivatives appearing in the differential equations, non-trivial denominator functions of the discrete representations for the derivatives, and the validity of special conditions holding for the differential equations and/or its solutions for the difference equation model and/or its solutions.

In order to be able to derive the denominator function ϕ the following consideration is made. If the compartment V and transitions to and from V are omitted in the $SVIHCDR$ model in Table 5.1, the $SIHCDR$ model is obtained. It is defined that $\tilde{N} = N - D = S + I + H + C + R$, and an actually negligible recruitment rate μ is added to the system [279]. Adding the differential equations of the $SIHCDR$ model yields the differential equation

$$\frac{d\tilde{N}(t)}{dt} = \mu \cdot \left(1 - \tilde{N}(t)\right), \tag{6.5}$$

that is solved by

$$\tilde{N}(t) = 1 + \left(\tilde{N}^0 - 1\right) \cdot e^{-\mu \cdot t} = \tilde{N}^0 + (N^0 - 1) \cdot (e^{-\mu \cdot t} - 1).$$

$$\text{with } \tilde{N}^0 = S(0) + I(0) + H(0) + C(0) + R(0). \tag{6.6}$$

It is defined that $\theta_I^n(t) = \beta^n \cdot \gamma^n(t) \cdot \left(1 - q^n(t)\right)$ is the transmission rate in the nth step.

With the aid of this and the denominator function, the $NSFD$ scheme can be established, which is provided in Table 6.4. It can be noticed that a standard incidence is used here.

Table 6.4 Implicit nonstandard finite difference scheme for the $SIHCDR$ model

$$\frac{S^{n+1} - S^n}{\phi(h)} = -\theta_I^n(t) \cdot (\varepsilon_I \cdot I^n + \varepsilon_H \cdot H^n) \cdot \frac{S^{n+1}}{N^{n+1} - D^{n+1}} - \mu \cdot S^{n+1} \ ,$$

$$\frac{I^{n+1} - I^n}{\phi(h)} = \theta_I^n(t) \cdot (\varepsilon_I \cdot I^{n+1} + \varepsilon_H \cdot H^n) \cdot \frac{S^{n+1}}{N^{n+1} - D^{n+1}} - (\omega_2^I + \eta^I + \mu) \cdot I^{n+1} \ ,$$

$$\frac{H^{n+1} - H^n}{\phi(h)} = \eta^I \cdot I^{n+1} - (\omega_3 + \lambda_1 + \xi + \mu) \cdot H^{n+1} \ ,$$

$$\frac{C^{n+1} - C^n}{\phi(h)} = \xi \cdot H^{n+1} - (\omega_4 + \lambda_2 + \mu) \cdot C^{n+1} \ ,$$

$$\frac{D^{n+1} - D^n}{\phi(h)} = \lambda_1 \cdot H^{n+1} + \lambda_2 \cdot C^{n+1} \ ,$$

$$\frac{R^{n+1} - R^n}{\phi(h)} = \omega_2^I \cdot I^{n+1} + \omega_3 \cdot H^{n+1} + \omega_4 \cdot C^{n+1} - \mu \cdot R^{n+1} \ .$$

Similar to the case of the continuous model deduced from Table 5.1, adding the equations in Table 6.4 yields

$$\frac{\tilde{N}^{n+1} - \tilde{N}^n}{\phi(h)} = \mu \cdot \left(1 - \tilde{N}^{n+1}\right). \tag{6.7}$$

The denominator function can be derived by comparing Equation (6.7) with the discrete version of Equation (6.6), that is

$$\tilde{N}^{n+1} = \tilde{N}^n + (\tilde{N}^n - 1) \cdot (e^{-\mu \cdot t} - 1), h = \Delta t, \tag{6.8}$$

such that the denominator function is defined by

$$\phi(h) = \frac{e^{-\mu \cdot h} - 1}{-\mu} = \frac{1 - \mu \cdot h + \frac{1}{2} \cdot \mu^2 \cdot h^2 + \ldots - 1}{-\mu} = h - \frac{1}{2} \cdot \mu \cdot h^2 = h + \mathcal{O}(h^2) \ . \tag{6.9}$$

This means that the solution of Equation (6.8) is exactly the discrete version of Equation (6.6) if we use the denominator function given in (6.9). In other words, using the NSFD the long term behaviour of the total population is properly modelled. An even more accurate way to compute the denominator function would take into account the transition rate Υ_i at which the ith compartment is entered by individuals for all model compartments $\mathcal{K}_i, i = 1, 2, \ldots$ [280]. In this case the parameter μ occurring in the denominator function in Equation (6.9) would be replaced by a parameter T^*. T^* could be determined as the minimum of the inverse transition parameters:

$$T^* = \min_{i=1,2,\ldots} \left\{ \frac{1}{\Upsilon_i} \right\}.$$

The NSFD scheme in Table 6.4 is formally implicit because it uses a non-local approximation.

It can easily be rearranged to get a scheme that can be evaluated sequentially in an efficient explicit way, i.e. a solution of a nonlinear system is unnecessary. We state this rearranged NSFD scheme in Table 6.5.

Table 6.5 Explicit nonstandard finite difference scheme for the $SIHCDR$ model

$$S^{n+1} = \frac{S^n}{1 + \phi(h) \cdot \left(\mu + \theta_I^n(t) \cdot \left(\varepsilon_I \cdot I^n + \varepsilon_H \cdot H^n \right) \cdot \frac{1}{N^{n+1}+D^{n+1}} \right)}, \tag{6.10}$$

$$I^{n+1} = \frac{I^n + \phi(h) \cdot \theta_I^n(t) \cdot \left(\varepsilon_H \cdot H^n \right) \cdot \frac{S^{n+1}}{N^{n+1}+D^{n+1}}}{1 + \phi(h) \cdot \left(\left(\omega_2^I + \eta^I + \mu \right) - \theta_I^{n+1}(t) \cdot \varepsilon_I \cdot \frac{S^{n+1}}{N^{n+1}+D^{n+1}} \right)}, \tag{6.11}$$

$$H^{n+1} = \frac{H^n + \phi(h) \cdot \eta^I \cdot I^{n+1}}{1 + \phi(h) \cdot \left(\omega_3 + \lambda_1 + \xi + \mu \right)}, \tag{6.12}$$

$$C^{n+1} = \frac{C^n + \phi(h) \cdot \xi \cdot H^{n+1}}{1 + \phi(h) \cdot \left(\omega_4 + \lambda_2 + \mu \right)}, \tag{6.13}$$

$$D^{n+1} = D^n + \phi(h) \cdot \left(\lambda_1 \cdot H^{n+1} + \lambda_2 \cdot C^{n+1} \right), \tag{6.14}$$

$$R^{n+1} = \frac{R^n + \phi(h) \cdot \left(\omega_2^I \cdot I^{n+1} + \omega_3 \cdot H^{n+1} + \omega_4 \cdot C^{n+1} \right)}{1 + \phi(h) \cdot \mu}. \tag{6.15}$$

The NSFD scheme is *positive preserving* i.e. it always produces non-negative solutions since all parameters and the denominator function are non-negative. Thus negative values for the solution are avoided. Moreover, stability with respect to the maximum norm is ensured [280]. The system of ODEs given in Table 6.5 is implemented in $MATLAB$ as an alternative to the implementation of the system of ODEs of the continuous $SIVCDR$ model in order to be able to compare the results obtained from the application the NSFD scheme to those obtained from the usage of $ode45$.

6.5.2   Implementation of a Nonstandard Finite Difference Scheme for the $SIHCDR$ Model

In this subsection, the solutions obtained from using the built-in standard solver $ode45$ in the implementation of the $SIHCDR$ model with underlying German data are compared to the solutions obtained by the application of the nonstandard finite difference scheme given in Table 6.5.

The parameter K_2 was fixed to the value 0.077 for Germany and 0.061 for Sweden again. Creating the first scenario in Figure 6.34, the parameter β was varied in a small interval for Germany while the other estimated parameters were estimated within bounds. Figures 6.35 and 6.36 were realized by varying the quarantine and contact rate. A modification of ε_I was implemented for Sweden, which is shown in Figure 6.37.

Variation of β, Germany
With respect to this scenario, the bounds of the estimated parameters were set to $\varepsilon_I \in [0.1, 0.2]$, $\varepsilon_H \in [0.05, 0.15]$, $q_1 \in [0, 0.2]$, $\iota \in [0.1, 0.5]$, $M_\iota \in [0.4, 0.7]$ for Germany. The result of the modification of β was investigated. As opposed to the figures in Sections 6.3 and 6.4, the range from zero to 80 is shown on the x-axis in the following diagram. Therefore, a fourth wave of the pandemic occurring the the second half of the year 2022 becomes visible.

For all four graphs in Figure 6.34 the results were $\varepsilon_I = 0.2$, $\varepsilon_H = 0.15$, $q_1 \approx 0$, $\iota = 0.1$, $M_\iota = 0.4$.

The extrema of the curves resulting from the mentioned parameters bound setting were on a similar level between all realized program runs. It has to be mentioned that the values obtained for the estimated parameters differed between any two successive program runs if the upper bound of ε_I was set to the value 0.3 instead of 0.2. The respective resulting four graphs concerning different allocations of β reached from progressions with very high peaks of over 300,000 to graphs tending to zero. Furthermore, an increase in the boundaries of q_1 lead to a minimization of the estimated value of this parameter to its lower bound in all realized program runs. The outcome $q_1 = 0.2$ already resulted in curves reaching a small maximum (for instance 40,470 for $\beta = 0.0335$) in the calendar week 48 but tending to zero from then.

In Figure 6.34 it becomes obvious that the assignment $\beta = 0.032$ ($\beta = 0.0325$) effects a 1.45-times (1.20-times) as high local maximum in the third as fourth wave, whereas $\beta = 0.033$ results in almost equal sizes of the peaks reached in the third and fourth wave, and $\beta = 0.0335$ leads to a local maximum in the third wave that is 13% smaller than in the fourth wave. The fourth wave of the corona virus

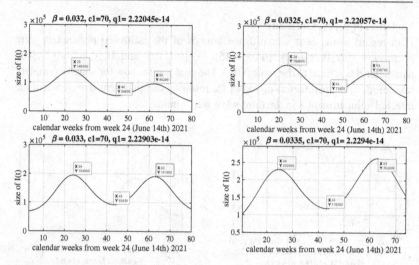

Figure 6.34 Prediction of the size of the infected compartment for Germany in the third wave, with $\beta \in [0.032, 0.0335]$

pandemic occurs from the calendar week 25 in 2022 and attains its peak in the calendar week 34 in 2022 in the upper left graph and calendar week 35 in 2022 in the other three graphs in Figure 6.34. The local maxima in the third wave are attained in the calendar week 47 in 2021 for $\beta = 0.032$ and calendar week 48 in 2021 for $\beta \in [0.0325, 0.0335]$. The variation of β in the interval $[0.032, 0.0335]$ does not have an as great influence as the modification of β in smaller intervals in Figures 6.11 to 6.14, where maxima were clearly reached earlier if the transmission risk was increased.

The average peak size in the given parameter scenario is 118,345 in the upper left, 150,750 in the upper right, 193,105 in the lower left and 246,800 in the lower right graph. Consequently, the peak growth in the regarded scenario is approximately 27.38% from the allocation $\beta = 0.032$ to $\beta = 0.0325$, 28.10% from $\beta = 0.0325$ to $\beta = 0.033$ and 27.81% from $\beta = 0.033$ to $\beta = 0.0335$.

It is also striking in the above figure that the assignment $\beta = 0.0335$ effects a comparatively large local minimum of $118,500$ infected individuals in the calendar week 25 in 2022. This minimum is more than twice as large as the minimal number of infected people achieved in the upper left graph, where $\beta = 0.032$.

Variation of q_1, c_1, Germany

With regard to the next scenario, the bounds of the estimated parameters were set to $\varepsilon_I \in [0.1, 0.2]$, $\varepsilon_H \in [0.05, 0.15]$, $\iota \in [0.1, 0.5]$ and $M_\iota \in [0.4, 0.7]$. The parameter β was fixed to the value 0.033. The goal was to show for which exemplary allocations of the parameters c_1 and q_1 the resulting compartment size curves of the infected compartment I in the third wave were similar.

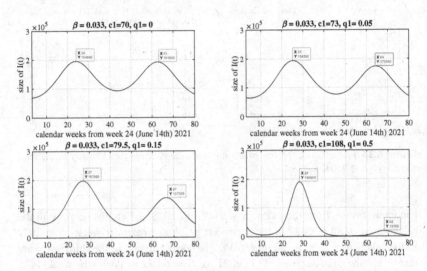

Figure 6.35 Prediction of the size of the infected compartment for Germany in the third wave, with $q_1 \in [0, 0.5]$, $c_1 \in [70, 108]$

For all of the plots in the above figure, resulting values were $\varepsilon_I = 0.2$, $\varepsilon_H = 0.15$, $\iota = 0.1$, $M_\iota = 0.4$.

In the 4 depicted graphs a maximal number of infected people of 190,600 to 197,600 in the third wave is reached. It can be said that a peak is achieved earlier in both regarded waves of the pandemic the smaller q_1 is here. This complies with the observations made in Figures 6.21 to 6.23.

What is most conspicuous in Figure 6.35 is the fact that the local maximum of the fourth wave is smaller the larger the parameter q_1 is chosen, even though the parameter c_1 is adjusted in the way that the peak sizes in the third wave are similar in all four graphs. The assignments $q_1 = 0$, $c_1 = 70$ lead to a local maximum of 191,600 in the calendar week 35, $q_1 = 0.05$, $c_1 = 73$ to a peak of in the calendar week 36, $q_1 = 0.15$, $c_1 = 79.5$ to a local maximum in the calendar week 39, and $q_1 = 0.5$, $c_1 = 108$ to a peak in the calendar week 40 in 2022.

The peaks in the third and fourth wave attain very similar sizes if q_1 is set to zero. This symbolizes a scenario of minimal intervention. The peak in the upper right graph is a 10.24% smaller than in the upper left graph ($q_1 = 0.05$, between minimal intervention and baseline scenario), but 26.47% larger than in the lower left graph (light lockdown measures). The local maximum of calendar week 40 in the lower right graph (lockdown scenario) is extremely small although the contact parameter c_1 is set to the value 108 here. Its size constitutes only 9.84% of the size of the peak in the upper left graph, where $c_1 = 70$. This big difference in the peak size as well as point in time of peak occurrence in the fourth wave of the pandemic is obviously accounted for by the difference in the size of the parameter q_1.

The correspondent progression of the compartment H is depicted below for the x-axis interval [0,40].

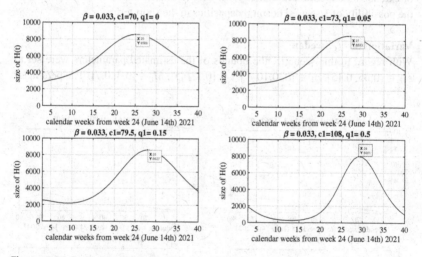

Figure 6.36 Prediction of the size of the hospitalized compartment for Germany in the third wave, with $q_1 \in [0, 0.5]$, $c_1 \in [70, 108]$

Figure 6.36 reveals that the size of the hospitalized compartment is the smaller the larger q_1 is between the calendar weeks 24 and 39. Whereas it is scarcely 3,000 in this parameter scenario if $c_1 = 70$, $q_1 = 0$ and has a tendency to increase from the beginning, it is close to 2,000 individuals if $c_1 = 108$, $q_1 = 0.5$ with a tendency to decrease like in Figure 6.35. The maximally reached size of the hospitalized compartment is very similar (8,001 to 8,627) in all four graphs in Figure 6.36 for the chosen allocations of the parameters c_1 and q_1. A local maximal size of the hospitalized compartment in the third wave is clearly achieved earlier for smaller q_1. It is reached in 4 weeks later in the case $c_1 = 108$, $q_1 = 0.5$ (calendar week 1 in 2022) than in the case $c_1 = 70$, $q_1 = 0$.

It is mentioned as an addition that the progression of the ICU compartment is similar to the course of the hospitalized compartment with respect to the examined parameter scenario. The size of the peak of the ICU compartment is predicted as lying between 598 (calendar week 2 in 2022) and 662 (calendar week 50 in 2021) individuals in four graphs corresponding to the same parameter allocations as in Figure 6.36. Moreover, the size of the deceased compartment increases by 37% in the case $q_1 = 0$, $c_1 = 70$ but 17% in the case $q_1 = 0.5$, $c_1 = 108$ in the regarded parameter scenario. Clearly, the information obtained from this and Figures 6.35 and 6.36 stresses the significance of quarantine in general. They highlight the relevance of extended intervention measures for a long-term attenuation of viral spread and the possibility of less used hospital capacities in the short term.

Variation of ε_I, Sweden

With regard to this scenario, the bounds of the estimated parameters were set to $\varepsilon_H \in [0.05, 0.15]$, $q_1 \in [0, 0.10^{-5}]$, $\iota \in [0.1, 0.5]$, $M_\iota \in [0.4, 0.7]$. The parameter

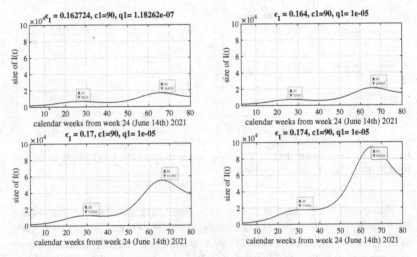

Figure 6.37 Prediction of the size of the infected compartment for Sweden in the third wave, with the lower bound of $\varepsilon_I \in [0.16, 0.174]$

β was fixed to the value 0.034. The consequence of the variation of the parameter ε_I was examined. The chosen x-axis interval is [0,80] again.

For all of the plots in the above figure, resulting estimated parameter sizes were $\varepsilon_H = 0.05$, $q_1 \approx 0$, $\iota = 0.1$, $M_\iota = 0.4$.

Figure 6.37 presents that the infected compartment comprises the more individuals the larger ε_I is in both regarded waves. The size of the peak reached in the third wave is more than 2.6 times larger for $\varepsilon_I = 0.174$ than $\varepsilon_I = 0.162724$ in the examined parameter scenario. The local maximum of the fourth wave is 5.56 times as large as for $\varepsilon_I = 0.174$ as for $\varepsilon_I = 0.162724$ here. In the case of $\varepsilon_I = 0.164$ ($\varepsilon_I = 0.174$) the increase originating from the choice of $\varepsilon_I = 0.162724$ in the compartment size is 24% (226.38%) with regard to the fourth and 10% (85%) with respect to the third wave. Thus the increase in the curve level in the fourth wave is generally larger than the one in the third wave.

The figure emphasizes the importance of isolation of infected cases, which effects a reduction of the size of the parameter ε_I owing to the reduced risk of transmission emerging from stricter compared to looser isolated individuals. In general, more asymptomatic cases might be prevented from infecting susceptibles by determining the amount of viruses in samples and so test for the degree of their infectiousness.

Markov Chain Epidemic Models

7

The ODE models described, established and implemented in Chapters 3 to 6 of this thesis are deterministic models. An alternative to deterministic approaches are stochastic compartment models. It can generally be said that stochastic models might be better suited than deterministic models in cases in which the number of infectious individuals is small or demographic/environmental variability could significantly impact the epidemic outcome [78]. Stochastic models possess some inherent randomness or uncertainty. Demographic stochasticity describes the randomness resulting from the inherently discrete nature of individuals with largest impact on small populations, whereas environmental stochasticity describes the randomness resulting from any change effecting an entire population [281]. The model also has to deal with uncertainty in data. An epidemic stochastic model can be a Markov population process.

Markov processes are the basis for general stochastic simulation methods known as *Markov Chain Monte Carlo* methods (MCMC). MCMC methods are Bayesian inferential methods, which make use of statistical inference and Bayes' theorem. Statistical inference is a statistical concept which draws conclusions with respect to hypotheses from observations. The Bayes' theorem describes the computation of conditional probabilities.

According to this mathematical theorem the probability of the occurrence of an event A under the condition that an event B occurred is given by [298]

$$\mathbb{P}(A|B) = \frac{\mathbb{P}(B|A) \cdot \mathbb{P}(A)}{\mathbb{P}(B)} .$$

In Markov processes, the so-called *Markov property* is assumed. This property is defined in Section 7.1. Methods of Bayesian inference are widely acknowledged nowadays since they do not only incorporate uncertainty in parameter values, but

© The Author(s), under exclusive license to Springer Fachmedien Wiesbaden GmbH, part of Springer Nature 2021
S. M. Treibert, *Mathematical Modelling and Nonstandard Schemes for the Corona Virus Pandemic*, BestMasters, https://doi.org/10.1007/978-3-658-35932-4_7

more importantly the randomness of the population size of infected hosts [282]. They are also frequently used in scientific approaches of modelling the spread of COVID-19, see for example [283–285].

Bayesian inference is often applied in so-called *multi-state models*, which can assume the Markov property and are introduced in Section 7.1. In Section 7.2, the Bayesian inference approach to compartment models is explained, which is useful for the derivation of the *Metropolis-Hastings algorithm*, which is a MCMC algorithm and described in Section 7.3.

7.1   Multi-state Models

A multi-state model is defined as a stochastic process which allows individuals to move within a finite discrete number of compartments or states, which could be clinical symptoms, biological markers, disease stages or disease recurrence in biomedical researches [286]. In medicine, these states can describe conditions like healthy, diseased and dead. Multi-state models can either be fitted assuming discrete or continuous time. Movement from one state to another is called a transition. States are called "transient" if a transition can emerge from the state or "absorbing" if no transition can emerge from the state [286]. So-called mortality models with only the two states "alive" and "dead", competing risk models extending the mortality model in the way that an individual may experience one of several failure outcomes, disease-progressive models, disability models or K-progressive models are examples of multi-state models [286]. The Markov property can but does not have to be assumed by a multi-state model.

Multi-state models are the most common models for describing the development of longitudinal failure time data. This is significant for modelling events with event-related dependencies like disease occurrence or risk of death [288]. Survival rates are an issue of interest in multi-state models and a survival analysis is usually connected to the application of a multi-state model.

For a probability space $(\Omega, \mathcal{F}, \mathbb{P})$ and a measurable space $(E, \mathcal{E})$, an E-valued stochastic process is a family of random variables $(T_n)_{N \in \mathbb{N}_0}$ [289, p. 2]. The mathematical definition of a multi-state model is the following:

A multi-state model is a stochastic process $\left(M(t), t \in [0, t_D], t_D < \infty \right)$ with a finite space $S = \{1, ..., S_{end}\}$, where t is some time and t_D is the time to death. Let F_t be a σ-algebra that is generated consisting of the observation of the process over the interval $[0, t]$ such as the states previously visited, transition times etc. A multi-state model is fully determined by certain transition-specific hazard rates. A *hazard rate function* $h(t)$ states how many objects fail, which can for instance mean die, on average per time interval. It is defined as

$$h(t) = \frac{\frac{d\mathcal{R}(t)}{dt}}{1 - \mathcal{R}(t)} = \frac{\mathcal{R}'(t)}{1 - \mathcal{R}(t)} = \frac{\mathcal{R}'(t)}{\mathcal{G}(t)},$$

where $\mathcal{G}$ is the survival function and $\mathcal{R}$ is the reliability function, which were defined in Subsection 4.2.1. The reliability function is also called the failure or lifetime distribution function. The hazard function can be described as the total number of failures within a population, divided by the total time expended by that population, during a particular measurement interval under stated conditions [290].

If F_t is the selected σ-algebra covering information over $[0, t]$, $(F_t)_t$ is a filtration, and $M_t, t \in [0, \infty)$ is a stochastic process, the hazard intensity of the next transition from a state $l \in \mathcal{S}$ to a state $m \in \mathcal{S}$ is defined by [291, p. 243]

$$h_{lm}(t) = \lim_{\Delta t \to 0} \frac{\mathbb{P}(M_{t+\Delta t} = m | F_t, M_t = l)}{\Delta t},$$

such that the total hazard out of state l is the sum over all states $m \neq l$:

$$h_l(t) = \sum_{m \neq l} h_{lm}(t).$$

Hazard rate functions provide the transition-specific hazards for movement from one discernible state to another. The transition intensity functions can also be used to compute the *mean sojourn time* [282]. A common assumption of multi-state models is time homogeneity, which means that the intensities are constant over time.

A way of dealing with multi-state models is to assume a homogeneous Markov model [291, p. 249], which is a Markov model in which the transition probabilities are time-independent. This type of model is usually used because of its simplicity [286]. Alternatively, multi-state models can assume a semi-Markov process. In this case, the next future transition depends on both the currently occupied state and the time of entry into the current state. This approach is considered more flexible in most cases. Furthermore, non-Markovian processes can be assumed in multi-state models, whose implementation has been challenging until specific "Markov-free" estimators for transition probabilities were introduced in the last decade [286].

Markov models are stochastic models, which are used to model systems that arbitrarily change, for which the Markov property holds. If $(E, \mathcal{E})$ is a measurable space, a Markov chain $M = (M_z)_{z \in \mathbb{N}_0}$ of order n is a stochastic process owning the Markov property, which is defined by [289, p. 4]

$$\mathbb{P}(M_{z+1} = s_{z+1}|M_0 = s_0, ..., M_z = s_z)$$
$$= \mathbb{P}(M_{z+1} = s_{z+1}|M_{z-n+1} = s_{z-n+1}, ..., M_z = s_z)$$

$$\forall z \in \mathbb{N}_0, s_0, ..., s_{z+1} \in E.$$

Here the future state depends on the n previous states. A Markov chain of order one is a Markov chain in which the future state depends on the current state only:

$$\mathbb{P}(M_{z+1} = s_{z+1}|M_0 = s_0, ..., M_z = s_z)$$
$$= \mathbb{P}(M_{z+1} = s_{z+1}|M_z = s_z) \quad \forall z \in \mathbb{N}_0, s_0, ..., s_{z+1} \in E .$$

An epidemic can be considered as a particular Markov process or chain with a block-structured generator in which the levels and phases correspond to the numbers of individuals per compartment.

Markov models provide a convenient framework for analysing structural mechanisms underlying social change and for extrapolating shifts in the state distribution of a population [292].

A Markov transition probability $\varphi_{lm}(v, t)$ for the transition from a state $l \in S$ occupied at time $v \in [0, t_D]$ to $m \in S$ occupied at time $t \in [0, t_D]$ is defined by [291, p. 244]

$$\varphi_{lm}(v, t) = \mathbb{P}(M_t = m|M_v = l), \quad v \leq t .$$

In Markov models, the transition probabilities can be computed from the intensities by solving the so-called forward *Kolmogorov differential equations* (Fokker-Planck equations) [282]. A reversible Markov chain owns the property that

$$\varphi_{lm}(v, t) \cdot \mathbb{P}(M_v) = \varphi_{lm}(t, v) \cdot \mathbb{P}(M_t) .$$

This feature is also called *detailed balance*.

It should be noted that apart from stochastic continuous epidemic models, in which the transitions can occur at any time point $t \in [0, \infty)$, discrete Markov chain models exist, in which the transitions occur at a fixed time $t \in \{0, \Delta t, 2\Delta t, ...\}$, for which the Markov transition probability can be rewritten as

$$\varphi_{lm}(t + \Delta t, t) = \mathbb{P}\Big(M(t + \Delta t) = m|M(t) = l\Big) .$$

When the transition probability $\varphi_{lm}(t + \Delta t, t)$ does not depend on t, the process is said to be *time homogeneous*. For stochastic compartment models, the process is time homogeneous because the deterministic model is autonomous [287, p. 86].

While the transition rates are usually standard estimates in continuous epidemic models, they are often reported in discrete epidemic models [286].

In the Metropolis Hastings algorithm stated in Section 7.3, Markov chains are used, that exhibit certain Markov transition probabilities. Transition probabilities can be regarded and are denoted as "samples" in the following sections.

7.2 The Bayesian Inference Approach to Compartment Models

Bayesian inference is a statistical framework for the estimation of parameters through observed data.

It allows for the realization of statistical inference based on Bayes' theorem, which expresses the relation between the updated knowledge called "the posterior", the prior knowledge and the knowledge coming from the observation expressed through the likelihood [293]. In this section, we let $D_j^{(i)}$, Φ and $\hat{\Phi}$ be defined as in Section 7.1. In a Bayesian framework the probability of an observed data set $\hat{\Phi}$ given a vector ϑ of sought parameters is the so-called *sampling density* or *sampling distribution* $\mathbb{P}(\hat{\Phi}|\vartheta)$. Differently expressed, it is the distribution of the observed data conditional on its parameters. It is a probability model for $\hat{\Phi}$ given the (partly) unknown parameter ϑ, that is randomly distributed according to a so-called prior distribution $\mathbb{P}(\vartheta)$ [283]. The prior distribution is the distribution of the parameters before any data is observed. Hence a prior distribution represents the information available before any data reports are known. It expresses the degree of uncertainty with respect to the parameter ϑ before a statistical analysis is implemented [294, p. 478], .

Let $f\left(D_j^{(i)}|\vartheta\right)$ be the probability of observing the measured size of the compartment $\mathcal{K}_i, i \in \{1, ..., s\}$ and $f\left(\Phi(t_j)|\vartheta\right)$ the probability of observing $\Phi(t_j)$ at the time instant $t_j, j \in \{1, ..., l\}$ for given ϑ.

The sampling density function $\mathbb{P}(\hat{\Phi}|\vartheta)$ describes the density of the conditional distribution of the sampled data $\hat{\Phi}$ for given parameters ϑ. It is proportional to the likelihood $L(\vartheta)$ and defined as [294, p. 478]

$$\mathbb{P}(\hat{\Phi}|\vartheta) = C \cdot L(\vartheta) = C \cdot f\left(\Phi(t_1)|\vartheta\right) \cdot f\left(\Phi(t_2)|\vartheta\right) \cdot ... \cdot f\left(\Phi(t_l)|\vartheta\right),$$

where C is a constant. The posterior distribution $\mathbb{P}(\vartheta|\hat{\Phi})$ of the parameter ϑ with its continuous parameter space Ω given data $\hat{\Phi}$ shall be computed. It is defined in accordance with Bayes' Theorem and given in Equation (7.1).

$$\mathbb{P}(\vartheta|\hat{\Phi}) = \frac{\mathbb{P}(\hat{\Phi}|\vartheta)\mathbb{P}(\vartheta)}{\mathbb{P}(\hat{\Phi})} = \frac{\mathbb{P}(\hat{\Phi}|\vartheta)\mathbb{P}(\vartheta)}{\int_\Omega \mathbb{P}(\hat{\Phi}|\vartheta)\mathbb{P}(\vartheta)\,d\vartheta} \propto L(\vartheta)\mathbb{P}(\vartheta) =: \pi(\vartheta|\hat{\Phi})\,.$$

$$(7.1)$$

It can be seen in the above equation that the posterior distribution is proportional to the unnormalized posterior distribution $\pi(\vartheta|\hat{\Phi})$, that is defined as the likelihood function of ϑ multiplied by the prior distribution of ϑ. The integral in the denominator symbolizes the necessary normalization constant such that the posterior distribution is indeed a proper distribution [295, p. 20]. The posterior distribution is the central quantity in Bayesian inference as it provides the probability distribution of the parameter after observing the data [295, p. 20]. The approximation of the posterior distribution given in Equation (7.1) is normally necessary as exact Bayesian inference is usually impossible.

Consequently, the determination of the likelihood of the parameter vector enables the obtainment of its unnormalized posterior distribution, which is also called the *target distribution*.

In the sequel, the probability density function (PDF) of a random variable X representing compartment sizes at the point x is denoted by $f(x)$ and the correspondent PDF depending on a parameter (vector) $\tilde{\vartheta}$ is denoted by $f_{\tilde{\vartheta}}(x)$.

The likelihood function of a parameter $\tilde{\vartheta}$ given the outcome x of a random variable X is defined in Equation (7.2).

$$\mathbb{P}_{\tilde{\vartheta}}(X = x) = f_{\tilde{\vartheta}}(x) = L(\tilde{\vartheta}|x) \qquad (7.2)$$

Let $X_1, ..., X_l$ be an available independent and identically distributed (i.i.d.) sample of l continuous random variables with realized values $\Phi_j := \Phi(t_j)$, $j \in \{1, ..., l\}$ corresponding to measured time series data.

Moreover, let X_j be a vector with variables adopting the values that represent the compartment sizes of the s compartments of a compartment model at the time instant t_j. Let the sought parameter vector ϑ be unknown.

The value of the joint probability density as a function of ϑ is called the likelihood of ϑ [294, p. 467],:

$$L(\vartheta) = f(X_1 = \Phi_1, ..., X_l = \Phi_l|\vartheta) \overset{X_j \text{ i.i.d}}{=} f(\Phi_1|\vartheta) \cdot ... \cdot f(\Phi_l|\vartheta) =: L_1(\vartheta) \cdot ... \cdot L_l(\vartheta)\,,$$

where $L_j(\vartheta)$ is the likelihood function associated with the jth independent data set, $j \in \{1, ..., l\}$. With the aid of the PDF $f(D_j^{(i)}|\vartheta)$ of the measured data concerning the ith compartment and j^{th} time series data set, it is defined by

$$L_j(\vartheta) = f(\Phi_j|\vartheta) := \prod_{i=1}^{s} f(D_j{}^{(i)}|\vartheta) \, .$$

Moreover, the respective Log-Likelihood function is given by

$$l(\vartheta) = \ln\Big(L(\vartheta)\Big) = \ln\Big(f(\Phi_1|\vartheta) \cdot \ldots \cdot f(\Phi_l|\vartheta)\Big) = \sum_{j=1}^{l} \ln\Big(f(\Phi_j|\vartheta)\Big) \, .$$

The Log-likelihood function is often preferred to the likelihood function because a sum appears in the used likelihood expression instead of a product. A sum is easier to optimize, since the derivative of a sum is the sum of derivatives, and very small (large) values in the case of discrete (continuous) random variables can be avoided due to the replacement of the product.

The distribution of the observations over time for the data sets of all observed points in time must be selected before fitting a system of ODEs $x'(t) = F(t, x, \vartheta)$ to the reported data. Common probability distributions used for this purpose comprise the normal distribution, Poisson distribution, and negative binomial distribution [283]. The density function of the Poisson distribution is used to account for noise in the data in the example below.

Let $D_j{}^{(i)}$ be described by the Poisson distribution with mean $\mu_j{}^{(i)}$ with regard to the following example. The expected value $\mathbb{E}(D_j{}^{(i)}) = \mu_j{}^{(i)}$ of the Poisson distribution changes depending on $t_j, j \in \{1, ..., l\}$. The PDF of the Poisson distribution is then defined as

$$f(D_j{}^{(i)}) = \frac{\exp(-\mu_j{}^{(i)}) \cdot \mu_j{}^{(i)}{}^{D_j{}^{(i)}}}{D_j{}^{(i)}!} \, .$$

Subsequently, the respective likelihood function is represented by

$$L(\vartheta) = \prod_{j=1}^{l} \prod_{i=1}^{s} \frac{\exp(-\mu_j{}^{(i)}) \cdot \mu_j{}^{(i)}{}^{D_j{}^{(i)}}}{D_j{}^{(i)}!}$$

$$= \frac{1}{D_1{}^{(1)}! \cdot \ldots \cdot D_1{}^{(s)}!} \cdot \ldots \cdot \frac{1}{D_l{}^{(1)}! \cdot \ldots \cdot D_l{}^{(s)}!}$$

$$\cdot \exp\Big(-\sum_{j=1}^{l} \sum_{i=1}^{s} \mu_j{}^{(i)}\Big) \prod_{j=1}^{l} \Big(\mu_j{}^{(1)}{}^{D_j{}^{(1)}} \cdot \ldots \cdot \mu_j{}^{(N)}{}^{D_j{}^{(s)}}\Big) \, .$$

The *Maximum-Likelihood* (ML) method is one of the most commonly used estimators in statistics. It differs from MCMC methods in the way that it is a non-Bayesian approach. Instead of applying Bayes' Theorem to determine a posterior distribution of the sought model parameter vector ϑ, ϑ is varied in the way that its likelihood function is maximized. The target of ML is to find the maximum possible product of the PDFs of the data points Φ_j, $j \in \{1, ..., l\}$, by variation of ϑ:

$$\hat{\vartheta}_{\text{ML}} = \underset{\vartheta \in \mathbb{R}^m}{\text{argmax}} \; L(\vartheta) = \underset{\vartheta \in \mathbb{R}^m}{\text{argmax}} \; \prod_{j=1}^{l} f(\Phi_j | \vartheta) \,.$$

Let $\hat{\vartheta}_{\text{ML}}$ be the value obtained from ML. Let there be m_1 parameters to be estimated in the model, that are comprised in a vector $\vartheta_1 \in \mathbb{R}^{m_1}$, and l points in time, at which data Φ_j, $j \in \{1, ..., l\}$, are observed.

The *Akaike Information Criterion* (AIC) value is defined as [283]

$$AIC = -2 \cdot \ln\!\left(L(\hat{\vartheta}_{\text{ML}})\right) + 2 \cdot m_1 \,.$$

If $m_1 > \frac{l}{40}$ the following corrected criterion AIC_C value should be used instead [283]:

$$AIC_C = AIC + \frac{2 \cdot m_1 \cdot (m_1 + 1)}{l - m_1 - 1} \,.$$

The AIC is useful in the selection of the model, with the aid of which the parameter values in ϑ are estimated. The principle of parsimony can be applied in the compartment model selection. A parsimonious model is the simplest model with the least assumptions and variables but with the greatest explanatory power for the disease process represented by the data [296]. The model selection method using the AIC considers the fit of the model to the data as well as the principle of parsimony.

The difference Δ_k between the AIC value AIC_k of the kth model and the minimum of all considered AIC values is

$$\Delta_k = AIC_k - \min_k AIC_k \,,$$

and measures the loss of information if model k is used instead of the model with the minimum AIC value [283]. If there are multiple possible models $\mathcal{M}_k$, $k = 1, 2, 3, ...$, the model $\mathcal{M}_k$ with the smallest value Δ_k should be selected. In the following section, the procedure of the Metropolis-Hastings algorithm is explained as an example of the class of MCMC algorithms.

7.3 The Metropolis-Hastings Algorithm

As explained in Section 7.2, Markov Chain Monte Carlo (MCMC) algorithms are a class of algorithms that are used to approximate a posterior distribution of parameters by randomly sampling the parameter space based on PDFs [283]. Instead of directly capturing an independent sample from the posterior distribution, a Markov chain is generated, whose iterations of the transition kernel converge to the posterior distribution.

The Markov chain generated by MCMC has the target distribution as its stationary distribution. After a certain number of steps and convergence time also called the *burn-in period* a sample which can be regarded as a sample from the posterior distribution is generated [294, p. 482]. The basic characteristic of MCMC sampling is that each sampled value depends (only) on the value sampled immediately prior (the Markov property) [297].

The MCMC procedure is based on the building of a Markov chain, whose stationary distribution is sought in order to sample from it.

The Metropolis Hastings algorithm is presented here because it is a classical MCMC method, which only requires the desired density to be known up to proportionality. In this algorithm, a new vector of parameter values $\vartheta^{(t)}$ is sampled iteratively, starting at a chosen value $\vartheta^{(0)}$. Per iteration, a new sample Ψ is drawn from a *proposal distribution* $f(\vartheta|\vartheta^{(t-1)})$, based on the previous vector $\vartheta^{(t-1)}$.

A value Ψ is accepted as the new $\vartheta^{(t)}$ with a certain probability α given in Equation (7.3), which is also called the *acceptance ratio* [298]:

$$\alpha(\Psi|\vartheta^{(t-1)}) = \min\left\{1, \frac{f(\Psi|\vartheta) \cdot \pi(\vartheta^{(t-1)}|\Psi))}{f(\vartheta^{(t-1)}|\vartheta) \cdot \pi(\vartheta^{(t-1)}|\Psi))}\right\} \overset{f \text{ symmetric}}{=} \min\left\{1, \frac{f(\Psi|\vartheta)}{f(\vartheta^{(t-1)}|\vartheta)}\right\}. \tag{7.3}$$

If Ψ is not accepted, then $\vartheta^{(t)}$ is set to the value $\vartheta^{(t-1)}$.

This process is continued until a sample path ("chain") has arrived at a stationary process and produces the target unnormalized posterior distribution. The choice of the proposal distribution is generally arbitrary, but it should be possible to easily generate random numbers from this distribution [294, p. 483].

The fulfilment of the inequation $f(\Psi) \geq f(\vartheta^{(t-1)})$ indicates that $f(\vartheta)$ has a low density near Ψ. An advantage of the Metropolis Hastings algorithm to other random number samplers is that the normalization constant of the posterior distribution does not have to be known because $f(\vartheta|\Psi)$ is considered in α in the ratio $\frac{f(\Psi|\vartheta)}{f(\vartheta^{(t-1)})}$ [294, p. 483].

The construction of the acceptance ratio guarantees that the stationary distribution of the resulting chain is the posterior distribution. It can be observed that the acceptance ratio increases for parameters yielding a higher posterior density such that the chain primarily moves along high posterior regions [295, p. 53].

The choice of the proposal distribution is a general issue with respect to the Metropolis Hastings algorithm, since state-independent proposal densities often lead to getting stuck at high posterior density points. For this reason, random walks are often used, which means that symmetric proposal distributions centred around the current parameter vector $\vartheta^{(t)}$ are multivariate normal or uniform distributions with mean $\vartheta^{(t)}$ [295, p. 54]. Advantages of such random walk proposals are the dependence of the acceptance probability on only the posterior values at the current proposed vector, or the possibility of achieving a high acceptance probability by choosing the variance of the proposal distribution sufficiently small [295, p. 54]. Nevertheless, too small variances can result in very small step sizes leading to very slow movement of the Markov chain.

An adaptive Metropolis algorithm was developed by Haario et al. [299]. Here, the transition proposal distribution is adjusted on the run according to the so far generated samples. The Gibbs Sampling is a special case of the Metropolis Hastings algorithm, which generates Markov chains by alternately drawing from distributions conditional to the current values of the remaining parameters, and updating the respective block of the drawn parameter [295, p. 55]. It should be kept in mind that the required conditional distributions are rarely given as even the full unnormalized posterior is not often available in closed-form in the case of an ODE model. A MCMC algorithm that can be shown to perform better than various other MCMC algorithms is the affine invariant ensemble MCMC algorithm. It uses K so-called walkers, whose positions are updated based on the present positions of the K walkers [293]. Another MCMC procedure is Slice Sampling, that has the objective of producing large step sizes and guaranteed acceptance by first uniformly sampling an unnormalized posterior density value and then uniformly sampling from the region with unnormalized posterior density [295, p. 55]. Further MCMC methods are proposed in the literature.

Résumé

8.1 Conclusion

In this thesis a compartment model named *SARS-CoV-2-fitted model* was established to represent the population dynamics experienced during the COVID-19 pandemic and serve as a convertible instrument for further analysis and predictions of the spread of SARS-CoV-2.

The deterministic SARS-CoV-2-fitted compartment model consists of a susceptible, a quarantine, an *all-or-nothing*-vaccinated, a confirmed exposed, an unconfirmed exposed, a confirmed asymptomatic infectious, an unconfirmed asymptomatic infectious, a confirmed symptomatic infectious, an unconfirmed symptomatic infectious, a hospitalized, an intensive care unit, a deceased and a recovered compartment. These 13 compartments were selected as they collectively enable the involvement of the latent period and the incubation period as indicators of the beginning of infectiousness or symptoms, a stage of isolation, the dark figure as the total number of unconfirmed cases, hospital sojourn times, disease-induced deaths, short-time exclusion from susceptibility through quarantine and long–time exclusion from susceptibility through effective vaccines. The most significant fact with regard to the model is its representation of stages of disease progression of distinct gravity.

Apart from the system of ordinary differential equations (ODEs) expressing the compartmental system, the first main contribution of this thesis are the characteristics of the 13 model compartments, which were described in detail with respect to their state of infection, relevant transition behaviours and relations to the other included compartments. This was accompanied by the definition of a rate of transition between each pair of adjacent compartments with exchange or one-sided transition of individuals. Assuming an exponentially distributed time to leave each

S. M. Treibert, *Mathematical Modelling and Nonstandard Schemes for the Corona Virus Pandemic*, BestMasters, https://doi.org/10.1007/978-3-658-35932-4_8

compartment, the transition rates are primarily based on the average period of stay in the compartment that is left by the individuals. An alternative to exponential modelling of the period of sojourn in infected compartments was given by the explanation of how the incubation period could be modelled as Erlang-distributed and how the serial interval could be modelled as hypoexponentially distributed.

It was also detailed in which ways the transmission risk, the contact rate and the quarantine rate, that all influence the transmission rates defined for the model, could be modelled as exponential or trigonometric functions for the purpose of depicting seasonal fluctuations of viral transmissibility or intervention measures already in values taken by certain model parameters. The possibility of introducing time delay to the system of differential equations was explained as an addition, which enables the inclusion of, for instance, the latent or incubation period in the transmission rate and is substantially relevant for future work.

Whereas several scientific papers focus on a specific feature of the spread of an infectious disease like COVID-19, as for example the prediction of the number of asymptomatic and other unconfirmed cases, the relations between different age groups in the population, the effect of future state interventions, the computation of the basic reproduction number R_0 etc. and create a deterministic or stochastic compartment model in view of a clear target, the SARS-CoV-2-fitted model and associated explanations serve as an enhanced framework, which can flexibly be specified with regard to the number and features of the included infected compartments or transition rates depending on the target of the corresponding model implementation. The explanations of multi-state and Markov epidemic models and Bayesian inference applied to compartment models also belong to this framework because they offer a competitive alternative to deterministic models and are the foundation of multiple realized implementations of stochastic compartment models used for forecasts of the COVID-19 spread across the world. Since an extensive framework for COVID-19 modelling was realized, the first objective of this thesis has been fulfilled.

Subsequent to the establishment of the model framework conditions, possible model variants were detailed and two of them were pictured along with their associated systems of ODEs and transition diagrams to adapt the SARS-CoV-2-fitted model to two different implementation targets as well as available underlying reported data sets. The 2 implemented variants were the $SIHCDR$ model with a pooled infected compartment comprising confirmed cases, and the $SVID$ age group model consisting of three different age groups and containing a vaccinated compartment. Thus two model specifications were realized. The predictions of future compartment sizes based on these two models is the second main contribution of this thesis.

The results of the $MATLAB$ implementation of the $SIHCDR$ model depicted in Figures 6.13 and 6.14 show that the presence of more and easier transmissible mutations, that effects an increased transmission risk β, already leads to a great increase in the number of weekly new infections if the increase in the size of β is only $5 \cdot 10^{-5}$ and thus 2–4 % of the value initially assigned to β. This emphasizes the necessity of containing the amount of transmissions with the aid of well-organized vaccination programs (best with all-or-nothing vaccines), extended testing not only of suspected cases, an effective quarantine of susceptibles and the isolation of infected cases.

The implementations in Section 6.3 imply that the development of the size of the infected compartment influences the progression of the size of the hospitalized and intensive care unit (ICU) compartments to a great extent. The rates of transition from the infected to the hospitalized and the hospitalized to the ICU compartment (η and ξ) are influenced by the case-hospitalization rate K and the portion ι of ICU admissions among hospitalizations. Efforts in the prevention of severe disease progressions, in particular through the early detection of infected cases and widespread vaccinations of the people in the elderly generation, would result in a reduction of both K and ι. Substantially in the case of a possible re-infection with a mutation after recovery from the infection with the original virus, efforts should be made to realize these measures.

Easier transmissible viral mutations indirectly result in more disease-induced deaths. If a mutated version of the virus is additionally associated with a generally higher mortality risk, the fast execution of a vaccination program is even more significant. The size of the rate of transition from the ICU compartment to the deceased compartment M_ι could be reduced by the deployment of enough ICU capacities in terms of personnel and beds for COVID-19 cases. The $MATLAB$ implementations reveal that an increased incidence owing to mutations can be effectively contained by reduced transmission modification factors i.e. a decrease in the risk emerging from the most infectious population groups (c.f. Figure 6.19). This could for instance be achieved through a comparatively stricter isolation of symptomatic cases. Maybe other factors that reduce transmission risks of infectious individuals will turn out to be efficient in the future. With regard to the SARS-CoV-2-fitted model, in which unconfirmed cases are included, extended testing would increase the number of case-confirmations and thus isolations of infected cases.

A vaccination program has to include available knowledge concerning the effects of accessible vaccines on the protection from the most aggressive mutations such that vaccines can be adapted to the behaviour of the mutated variants. The prioritization of the most affected population groups but exclusion of possible risks accompanying the vaccination of people is also important. The most significant target of nationwide vaccinations that are effective against mutations and prevent the transmission through vaccinated individuals is the high enough reduction of the transmission risk β to flatten the curve, as demonstrated in Figures 6.15 and 6.16.

The relation between the influences of the contact rate and the quarantine rate on the number of weekly newly confirmed infections are conveyed in Figures 6.20 to 6.29. Here, the implementations show that a decrease in the size of the contact rate parameter c_1 can result in a containment of high incidence caused by an decreased quarantine rate parameter q_1. This holds for the implementations based on both German and Swedish data. For example, the allocation $c_1 = 87$, that effects a maximal number of 97 and a minimal number of 77 total weekly contacts over the course of time, with an applied contact rate parameter $q_1 = 0.05$ (i.e. a scenario of minimal intervention with comparatively low contact rate) leads to a peak of a very similar height as the allocation $c_1 = 107.5$, $q_1 = 0.26$ (i.e. a lockdown scenario with an improbably high contact rate) for Sweden. In the more probable case of an increased (a decreased) contact rate associated with a simultaneously decreased (increased) quarantine rate, incidence would strongly increase (decrease). The calendar week in which an extremum is reached strongly depends on the size of q_1 here. The shifts of the contact rate and quarantine rate on x-axis have an even greater impact on the point in time at which a peak is attained.

The $MATLAB$ implementation of the $SVID$ age group model with a distinct exponential vaccination rate per age group in Section 6.4 reveals the force of nationwide vaccinations to continuously reduce the number of new infections over the whole observation period. It shows that a small increase in the exponent of the exponential function, i.e. from $\frac{1}{13}$ to $\frac{1}{8}$, yields a clearly visible decrease in the number of weekly newly confirmed infections.

Regarding the $MATLAB$ implementation of the nonstandard finite difference (NSFD) scheme in Section 6.5, an increased transmission risk causes a generally higher incidence level, but in particular a larger peak in the fourth wave in the year 2022. A size of the quarantine rate parameter q_1 corresponding to a lockdown scenario yields an extreme decrease in the number of newly confirmed infections in the fourth wave of the pandemic compared to a scenario with minimal or no intervention. Apart from this, the implementation of the NSFD scheme conveys that the number of German COVID-19 hospitalizations is on a very low level in the beginning of the third wave in the lockdown scenario, which more open hospital capacities at peak times are a consequence of. Thus the implementation suggests that a high quarantine rate is effective in containing the spread at least over short periods even if many contacts within the population cannot be prevented. As the application of the NSFD scheme for the $SIHCDR$ model leads to realistic results, and this method exhibits a number of favourable features like the preservation of positivity and correct long time behaviour, the performance of NSFD schemes applied to further compartment models should be examined.

Realistic results concerning future compartment sizes under different model parameter variations were achieved with the aid of implementing two compartment models emerging from the SARS-CoV-2-fitted model. One of these models was specified to forecasts of the infected, hospitalized, ICU and deceased class, and the other one to vaccination effects in an age-structured population. Beside that, two different methods of dealing with the systems of ODEs were applied, and the ℓ^2-error minimization in the implementations was sustained by some theory of nonlinear least squares. Subsequently, the second objective of this thesis has been fulfilled.

8.2   Future Work

Since we are observing a couple of delay mechanisms in the COVID-19 modelling, like delay in reporting confirmed cases, delayed hospitalization, waning immunity effects, *delay differential equations* (DDEs) could be included in the implementation, see [220] and the references therein. For example, the level of antibodies depending on the time lapsed can be modelled by a probability distribution DDE models are characterized by a higher degree of accuracy than SIR models. An aspect that could be considered in future work concerning the application of the SARS-CoV-fitted model to predict future compartment sizes is using statistical modelling in the form of regression models or deep learning techniques such as neural networks to handle missing data reports. Similar to DDEs, a statistical regression model building on differential equations and a nonlinear optimization problem can depend on points in time in the past.

Whereas the effects of generally present mutations compared to a scenario without any aggressive mutation can be included in the SARS-CoV-2-fitted model by adjusting the transmission rate, the impact of two or more different mutations and associated infection dynamics exerted on the population could be incorporated by creating a model and system of ODEs in which one or more separate infected compartments per mutated virus variant are included. A model with two mutations was realized in a scientific approach by Gonzalez-Parra et al. [301], whose model analysis revealed that even if the new variant has the same death rate, its high transmissibility can increase the number of infected people, those hospitalized, and deaths. Apart from this, a future implementation of the SARS-CoV-2-fitted model should involve leaky vaccinations with the aid of then potentially available reliable data concerning the waning effect of vaccine protection. This waning effect could be modelled with the aid of probability distributions just like the waning effect of the protection from transmitting the infection after symptom disappearance.

A very interesting question is the impact of efforts concerning the attainment of herd immunity on the size of the infected compartment and the deceased compartment in the presence of viral mutations. In future approaches, the effect of saturation of the need for vaccination (i.e the attainment of vaccinations for all individuals who want to be vaccinated) should be investigated in more detail. Vaccination strategies could be more precisely incorporated into the SARS-CoV-2-fitted model in future approaches by involving different vaccines and gaining more exact knowledge concerning the effects of accessible vaccines against mutations.

Apart from this, the optimal control of different factors influencing the size of the infected, hospitalized, ICU and deceased compartment should be regarded in terms of an optimization of the dynamic system. Here, *controls* have to be chosen optimally to optimize an objective function taking into account certain *constraints*. If a compartment model based on a system of ODEs is built upon, a deterministic system is given, whose evolution over time can be predicted knowing the initial state and control inputs. Constraints can for instance be the socio-economic costs effected by intervention measures or a limited number of intensive care capacities that may not be exceeded. The objective function contains costs as well as the targets of intervention measures. An objective could be to find an optimal strategy for releasing individuals from a quarantine compartment S_q or a compartment whose individuals experience non-pharmaceutical interventions (NPIs) to the class S and keep the number of infected individuals below a certain maximum value to prevent an exponential growth of infection numbers. Another objective could be to find an optimal strategy for minimizing treatment costs and NPIs leading to economic losses while minimizing the size of the infected compartment.

The next, straight forward step would be to consider the proposed system of ODEs for COVID-19 modelling on a graph network, as done by Brockmann [196, 197]. Here, the nodes may represent local populations as cities or countries. An edge weight may symbolize the relative frequency of contact between two individuals, the probability of a contact leading to a transmission may depend on the edge weights and a virulence factor, the basic reproduction number R_0 may be defined as the mean node degree, and vaccination may be included by removing a certain portion of nodes (all-or-nothing vaccine) or edges (leaky vaccine) from the network [302]. The effects of mobility restrictions or lockdown measures could be modelled by simply adjusting the corresponding transmission rates. So-called *global mobility network* approaches are applicable to SIR dynamics in order that the movement between different populations composed of susceptible, infected and recovered subpopulations can be described by an additional differential equation. Here, a global mobility rate can be defined as the rate to leave a node by a random individual [197]. Transition rates are then defined as conditional probabilities of randomly chosen individuals to move from one to another location within a time step, which a Markov transition matrix arises from [197]. With the aid of the *effective distance* introduced by Brockmann, the spreading speed and arrival times of an infectious disease as well as functional relationships between epidemiological and mobility parameters can be determined [196]. The effective distance between two local populations can be considered as the length of the path between the two locations, that is most likely selected by a random walking individual in the population among multiple selectable routes [303].

In extreme cases, the taken measure can be lead to break up chains of infection and isolate parts of the network. Consequently, the network is split into small clusters. In the same way, the spread of mutations, that exist initially only in a few nodes, to the whole network can be described. Doing so, the modelling of effects of travel activities can be included in a simple way.

Finally, the major future progress will consist in using PDEs for the disease modelling instead of comparatively simple ODEs. Doing so, the spatial dependence of the pandemic spread could be investigated in much more detail. Then, the network is defined by a metric graph, i.e. a graph where the connections between nodes (called *bonds*) are intervals and not only undirected links. The lockdown measures and mobility restrictions or effects of travel activities can then be modelled in a much more sophisticated way, by employing suitable, especially designed conditions at the branching points of the metric graph. These conditions model exactly the spatial effects of the restrictions or the relaxations, much better than simply adjusting transition rates as in the ODE case before. However, these branching points conditions are rather complicated and computationally expensive.

Bibliography

[1] World Health Organization: https://apps.who.int/iris/bitstream/handle/10665/332197/WHO-2019-nCoV-FAQ-Virus_origin-2020.1-eng.pdf.

[2] A. L. Rasmussen: 'On the origins of SARS-CoV-2' in Nature Medicine, Volume 27, Issue 9, January 2021.

[3] NDR: https://www.ndr.de/nachrichten/hamburg/Hamburger-Forscher-Coronavirus-stammt-wohl-aus-Labor,corona6764.html.

[4] European Centre for Disease Prevention and Control: https://www.ecdc.europa.eu/en/geographical-distribution-2019-ncov-cases, last access: April 5th 2021, 08:00 am.

[5] Johns Hopkins University: https://coronavirus.jhu.edu/data/new-cases, last access: February 16th 2021, 11:00 am.

[6] L. Bretschger, E. Grieg, P. J. J. Welfens, T. Xiong: 'COVID-19 Infections and Fatalities Developments: Empirical Evidence for OECD Countries and Newly Industrialized Economies' in International Economics and Economic Policy, Volume 17, pp. 801–147, August 2020.

[7] Statista.de: https://www.statista.com/statistics/1043366/novel-coronavirus-2019ncov-cases-worldwide-by-country/, last access: April 1st 2021, 09:50 am.

[8] Statista.de: https://www.statista.com/statistics/1110187/coronavirus-incidence-europe-by-country/, last access: April 1st 2021, 09:50 am.

[9] Verband der forschenden Pharma-Unternehmen: https://www.vfa.de/de/arzneimittel-forschung/woran-wir-forschen/impfstoffe-zum-schutz-vor-coronavirus-2019-ncov.

[10] Robert-Koch Institute: https://www.rki.de/DE/Content/InfAZ/N/Neuartiges_Coronavirus/Steckbrief.html, last access: February 16th 2021, 10:00 am.

[11] European Centre for Disease Prevention and Control https://www.ecdc.europa.eu/en/COVID-19/timeline-ecdc-response, last access: February 20th 2021, 09:00 am.

[12] Ourworldindata.org from Johns Hopkins University CSSE COVID-19 Data: https://ourworldindata.org/coronavirus.

[13] Ourworldindata.org: https://ourworldindata.org/grapher/continents-according-to-our-world-in-data.

[14] Robert-Koch Institute: https://www.rki.de/DE/Content/InfAZ/N/Neuartiges_Coronavirus/Daten/Altersverteilung.html, last access: January 20th 2021, 4:30 pm.

[15] Robert-Koch Institute: 'Seroepidemiologische Studie zur Verbreitung von SARS-CoV-2 in der Bevölkerung an besonders betroffenen Orten in Deutschland – Studienprotokoll

von Corona-Monitoring lokal' in 'Journal of Health Monitoring', Issue 5, pp. 4–5, August 2020.

[16]   K. Swinkels, G. Knight, F. Lakha, Q. Leclerc, A. Horn, N. Fuller: https://covid19settings.blogspot.com/p/about.html.

[17]   Wikipedia: https://en.wikipedia.org/wiki/Superspreading_event.

[18]   Deutsches Ärzteblatt: https://www.aerzteblatt.de/nachrichten/121000/Mehr-COVID-19-Infektionen-nach-Querdenken-Demonstrationen.

[19]   D. Majra, J. Benson, J. Pitts, J. Stebbing: 'SARS-CoV-2 (COVID-19) superspreader events' in Elsevier Public Health Emergency Collection, Volume 81, pp. 36–40, January 2021.

[20]   H. Streeck, B. Schulte, B. M. Kümmerer et al.: 'Infection fatality rate of SARS-CoV2 in a super-spreading event in Germany' in Nature Communications, November 2020.

[21]   K. A. Fisher, M. W. Tenforde, L. R. Feldstein et al.: 'Community and Close Contact Exposures Associated with COVID-19 Among Symptomatic Adults ≤ 18 Years in 11 Outpatient Health Care Facilities – United States, July 2020' in Morbidity and Mortality Weekly Report, Volume 69, pp. 1258–1264, September 2020.

[22]   Y. Furuse, E. Sando, N. Tsuchiya et al.: 'Clusters of Coronavirus Disease in Communities, Japan, January-April 2020' in Emerging Infectious Diseases, Volume 26, pp. 2176–2179, September 2020.

[23]   ZDF: https://zdfheute-stories-scroll.zdf.de/Corona_Cluster/index.html.

[24]   Robert-Koch   Institute: https://www.rki.de/DE/Content/InfAZ/N/Neuartiges_Coronavirus/Fallzahlen.html.

[25]   L. L. Ren, Y. M. Wang, Z. Q. Wu et al.: 'Identification of a novel coronavirus causing severe pneumonia in human: a descriptive study' in Chinese Medical Journal, Volume 133, Issue 9, pp. 1015–24, May 2020.

[26]   The British Journal of Medicine Best Practice: https://bestpractice.bmj.com/topics/en-us/3000168/aetiology#referencePop33.

[27]   World Health Organization: https://www.who.int/docs/default-source/coronaviruse/who-china-joint-mission-on-COVID-19-final-report.pdf?sfvrsn=fce87f4e_2.

[28]   B. Hu, H. Guo, P- Zhou, Z. Shi: 'Characteristics of SARS-CoV-2 and COVID-19' in Nature Review Microbiology, October 2020.

[29]   M. Gheblawi et al.: 'Angiotensin-Converting Enzyme 2: SARS-CoV-2 Receptor and Regulator of the Renin-Angiotensin System' in Public Health Emergency Collection, May 2020.

[30]   K. G. Andersen, A. Rambaut, W. I. Lipkin, E. C. Homes, R. F. Garry: 'The proximal origin of SARS-CoV-2' in Nature Medicine, Volume 26, pp. 450–452, March 2020.

[31]   N. Zhu et al.: 'A Novel Coronavirus from Patients with Pneumonia in China, 2019' in The New England Journal of Medicine, Volume 382, pp. 727–733, February 2020.

[32]   Springer   Medizin: https://www.springermedizin.de/differences-and-similarities-between-sars-cov-and-sars-cov-2-spi/18237540.

[33]   L. Lu, W- Zhong et al.: 'A comparison of mortality-related risk factors of COVID-19, SARS, and MERS: A systematic review and meta-analysis' in Journal of Infection, Volume 81, Issue 4, pp. 18–25, July 2020.

[34]   Robert-Koch Institute, Germany: https://behoerden.blog/wp-content/uploads/2020/05/Coronavirus-SARS-CoV-19-Steckbrief-RKI-7.5..pdf.

[35]   Y. Yang, W. Shang, X. Rao: 'Facing the Covid-19 outbreak: What should we know and what could we do?' in Journal of Medical Virology, Volume 92, pp. 536–537, April 2020.

[36]   World Health Organization: https://apps.who.int/iris/bitstream/handle/10665/112656/9789241507134_eng.pdf;jsessionid=41AA684FB64571CE8D8A453C4F2B2096?sequence=1.

[37]   M. Pfeifer, O. W. Hamer: 'COVID-19-Pneumonie' in Der Internist, Springer Nature, Volume 8, 2020.

[38]   Deutsches Ärzteblatt: https://www.aerzteblatt.de/nachrichten/114701/SARS-CoV-2-Evidenz-spricht-gegen-Ansteckung-ueber-die-Luft.

[39]   N. van Doremalen, T. Bushmaker, D. H. Morris et al.: 'Aerosol and Surface Stability of SARS-CoV-2 as Compared with SARS-CoV-1', The New England Journal of Medicine, Volume 382, pp. 1564–1567,April 2020.

[40]   S. Asadi et al.: 'Aerosol emission and superemission during human speech increase with voice loudness' in Scientific Reports, Volume 9, February 2019.

[41]   Y. Pan, D. Zhang et al.: 'Viral load of SARS-CoV-2 in clinical samples' in The Lancet Infectious Diseases, Volume 20, Issue 4, pp. 411–412, April 2020.

[42]   W. Wang et al.: 'Detection of SARS-CoV-2 in Different Types of Clinical Specimens', Jama Network, May 2020.

[43]   World Health Organization: 'Transmission of SARS-CoV-2: implications for infection prevention precautions', July 2020.

[44]   C. Lübbert, T. Grünewald, R. Gottschalk, R. Kurth, B. Ruf: 'Respiratorische Virusinfektionen: Klinische Differenzialdiagnose' in Deutsches Ärzteblatt, Volume 100, Issue 43, November 2003.

[45]   Deutsches Ärzteblatt: https://www.aerzteblatt.de/nachrichten/119475/Was-COVID-19-von-der-saisonalen-Influenza-unterscheidet.

[46]   Deutschlandfunk: https://www.deutschlandfunk.de/COVID-19-wie-verlaeuft-eine-infektion-mit-dem-coronavirus.2852.de.html?dram:article_id=473716.

[47]   Centers for Disease Control and Prevention: https://www.cdc.gov/training/QuickLearns/exposure/2.html.

[48]   P. Banka, C. Comiskey: 'The incubation period of COVID-19: A scoping review and meta-analysis to aid modelling and planning' in medRxiv – PreprintServer for Health Sciences, 20216143, November 2020.

[49]   H. Peckham, N. de Gruijter, C. Raine et al.: 'Male sex identified by global COVID-19 meta-analysis as a risk factor for death and ITU admission' in Nature Communications, December 2020.

[50]   K. S. Cheung, I. F. Hung, P. Y. Chan et al: 'Gastrointestinal Manifestations of SARS-CoV-2 Infection and Virus Load in Fecal Samples From a Hong Kong Cohort: Systematic Review and Meta-analysis' in Gastroenterology, Volume 159, Issue 1, pp. 81–95, July 2020.

[51]   M. Ackermann et al.: 'Pulmonary vascular endothelialitis, thrombosis, and angiogenesis in COVID-19' in The New England Journal of Medicine, Volume 383, pp. 120–128, July 2020.

[52]   M. Dreher, A. Kersten, J. Bickenbach et al.: 'Charakteristik von 50 hospitalisierten COVID-19-Patienten mit und ohne ARDS' in Deutsches Ärzteblatt, Volume 117, pp. 271–278, April 2020.

[53]   Y. Tu, R. Jennings, B. Hart et al.: 'Swabs Collected by Patients or Health Care Workers for SARS-CoV-2 Testing' in New England Journal of Medicine, Volume 383, Issue 5, pp. 494–496, July 2020.

[54]   A. W. Byrne, D. McEvoy, A. B. Collins et al.: 'Inferred duration of infectious period of SARS-CoV-2: rapid scoping review and analysis of available evidence for asymptomatic and symptomatic COVID-19 cases' in The British Journal of Medicine, Volume 10, Issue 8, July 2020.

[55]   C. Karagiannidis, C. Mosert, C. Hentschker et al.: 'Case characteristics, resource use, and outcomes of 10021 patients with COVID-19 admitted to 920 German hospitals: an observational study' in The Lancet Respiratory Medicine, Volume 8, Issue 9, July 2020.

[56]   European Centre for Disease Prevention and Control: https://www.ecdc.europa.eu/en/COVID-19/latest-evidence/clinical.

[57]   European Centre for Disease Prevention and Control: 'COVID-19 testing strategies and objectives', September 2020.

[58]   Centers for Disease Control and Prevention: https://www.cdc.gov/coronavirus/2019-ncov/symptoms-testing/symptoms.html.

[59]   K. Yuen, Z. Ye, C. Chan, D. Jin: 'SARS-CoV-2 and COVID-19: The most important research questions' in Cell and Bioscience, March 2020.

[60]   J. R. Lechien, C. M. Chiesa-Estomba, S. Place et al.: 'Clinical and epidemiological characteristics of 1420 European patients with mild-to-moderate coronavirus disease 2019' in Journal of Internal Medicine, Volume 288, Issue 3, pp. 335–344, September 2020.

[61]   J. J. Zhang, K. S. Lee, L. W. Ang et al.: 'Risk Factors for Severe Disease and Efficacy of Treatment in Patients Infected With COVID-19: A Systematic Review, Meta-Analysis, and Meta-Regression Analysis' in Clinical Infectious Diseases, Volume 71, Issue 16, pp. 2199–2206, November 2020.

[62]   A. B. Dochtery, E. M. Harrison, C. A. Green et al.: 'Features of 20133 UK patients in hospital with COVID-19 using the ISARIC WHO Clinical Characterisation Protocol: prospective observational cohort study' in The British Journal of Medicine, Volume 369, May 2020.

[63]   Z. Feng, Q- Yu et al.: 'Early prediction of disease progression in COVID-19 pneumonia patients with chest CT and clinical characteristics' in Nature Communications, October 2020.

[64]   Bundesgesundheitsministerium: https://www.bundesgesundheitsministerium.de/coronatest.html.

[65]   M. T. Osterholm, M. Olshaker: 'Let's get real about Coronavirus Tests' in The New York Times, April 2020.

[66]   R. L. Schlenger: 'PCR-Tests auf SARS-CoV-2: Ergebnisse richtig interpretieren' in Deutsches Ärzteblatt, Volume 117, Issue 24, June 2020.

[67]   I. Arevalo-Rodriguez et al.: 'False-negative results of initial RT-PCR assays for COVID-19: A systematic review' in PLOS ONE, December 2020.

[68]   J. Watson, J. E. Brush: 'Interpreting a COVID-19 test result' in The British Journal of Medicine, Volume 369, May 2020.

[69] Robert-Koch Institute: https://www.rki.de/DE/Content/InfAZ/N/Neuartiges_ Coronavirus/Vorl_Testung_nCoV.html;jsessionid=650B0885917709820CA82D0593 880F9B.internet081?nn=13490888.

[70] Robert-Koch Institute: https://www.rki.de/DE/Content/InfAZ/N/Neuartiges_ Coronavirus/Teststrategie/Nat-Teststrat.html.

[71] Zusammengegencorona: https://www.zusammengegencorona.de/testen/allgemeine-infos-zum-testen/.

[72] Robert-Koch Institute: https://www.rki.de/DE/Content/InfAZ/N/ Neuartiges_Coronavirus/Massnahmen_Verdachtsfall_Infografik_DINA3. pdf;jsessionid=763E872EE8335AE6462A630F36064F51.internet102?__ blob=publicationFile.

[73] Robert-Koch Institute: https://www.rki.de/DE/Content/InfAZ/N/Neuartiges_ Coronavirus/Vorl_Testung_nCoV.html.

[74] C. Drosten: 'Ein Plan für den Herbst' in DIE ZEIT, Volume 33, August 2020.

[75] Centers for Disease Control and Prevention: https://www.cdc.gov/nonpharmaceutical-interventions/index.html.

[76] S. Khailaie et al.: 'Development of the reproduction number from coronavirus SARS-CoV-2 case data in Germany and implications for political measures' in BMC Medicine, Volume 19, Article Number 32, Janaury 2021.

[77] M. V. Barbarossa, J. Fuhrmann: 'Compliance with NPIs and possible deleterious effects on mitigation of an epidemic outbreak' , in Preprints.org, 2021020178, February 2021.

[78] M. V. Barbarossa et al.: 'Modeling the spread of COVID-19 in Germany: Early assessment and possible scenarios' in PLoS One, 0238559, September 2020.

[79] J. Dehing et al.: 'Inferring change points in the spread of COVID-19 reveals the effectiveness of interventions' in Science, Volume 369, July 2020.

[80] Bundesgesundheitsministerium: https://www.bundesgesundheitsministerium.de/ coronavirus/chronik-coronavirus.html, last access on February 15th 2021, 12:10 pm.

[81] S. Imöhl, A. Ivanov: 'Coronavirus: So hat sich die Lungenkrankheit in Deutschland entwickelt', Article in Handelsblatt, February 2021.

[82] European Centre for Disease Prevention and Control: https://www.ecdc.europa.eu/ sites/default/files/documents/COVID-19-vaccination-and-prioritisation-strategies. pdf.

[83] A. Kavaliunas, P. Ocaya, J. Mumper, I. Lindfeldt, M. Kyhlstedt: 'Swedish policy analysis for COVID-19' in Elsevier Public Health Emergency Collection, Volume 9, pp. 598–612, August 2020.

[84] Government Offices of Sweden: https://www.government.se/government-policy/the-governments-work-in-response-to-the-virus-responsible-for-covid-1/.

[85] J. Vlachos, E. Hertegård, H. Svaleryd: 'School closures and SARS-CoV-2. Evidence from Sweden's partial school closure' in medRxiv – PreprintServer for Health Sciences, 20211359, December 2020.

[86] J. F. Ludvigsson: 'Children are unlikely to be the main drivers of the Covid-19 pandemic – a systematic review' in Wiley Public Health Emergency Collection, May 2020.

[87] M. Paterlii: 'Closing borders is ridiculous': the epidemiologist behind Sweden's controversial coronavirus strategy', Article in Nature, April 2020.

[88] R. Weichert: 'Die Kritik prasselt weiter auf Anders Tegnell ein' in Stern, January 2021.

[89]   Frankfurter Allgemeine Zeitung (FAZ): https://www.faz.net/aktuell/politik/ausland/
        schwedens-koenig-kampf-gegen-corona-pandemie-misslungen-17106735.html.

[90]   Ourworldindata.org: https://ourworldindata.org/coronavirus-data-
        explorer?zoomToSelection=true&minPopulationFilter=1000000&
        country=GBR~ITA~DEU~SWE&region=World&deathsMetric=true&
        interval=smoothed&hideControls=true&perCapita=true&smoothing=7&
        pickerMetric=location&pickerSort=asc

[91]   Public   Health   Agency   of   Sweden: https://www.folkhalsomyndigheten.se/
        contentassets/53c0dc391be54f5d959ead9131edb771/infection-fatality-rate-COVID-
        19-stockholm-technical-report.pdf.

[92]   Covid19insweden.com: https://www.covid19insweden.com/en/.

[93]   H. Webster, P. Phillips, Oxford COVID-19 Government Response Tracker: https://
        www.bsg.ox.ac.uk/research/research-projects/coronavirus-government-response-
        tracker#data.

[94]   Ourworldindata.org: https://ourworldindata.org/grapher/covid-stringency-index?
        stackMode=absolute&time=2020-05-07&region=World.

[95]   Ourworldindata.org: https://ourworldindata.org/coronavirus/country/germany?
        country=~DEU#government-stringency-index.

[96]   Paul-Ehrlich Institute: https://www.pei.de/DE/arzneimittel/impfstoffe/COVID-19/
        COVID-19-node.html, last access on February 15th 2021, 3:00 pm.

[97]   Centers for Disease Control and Prevention: https://www.cdc.gov/coronavirus/2019-
        ncov/vaccines/different-vaccines/mRNA.html.

[98]   Biontech: https://biontech.de/COVID-19-portal/mrna-vaccines.

[99]   J. A. de Soto: 'Evaluation of the Moderna, Pfizer/BioNtech, AstraZeneca/Oxford and
        Sputnik V Vaccines for COVID-19' in OsfPreprints, December 2020.

[100]   Robert-Koch   Institute: https://www.rki.de/DE/Content/Kommissionen/STIKO/
        Empfehlungen/Vierte_Empfehlung_2021-04-01.html   https://www.rki.de/
        SharedDocs/FAQ/COVID-Impfen/gesamt.html.

[101]   RND: https://www.rnd.de/politik/stiko-empfiehlt-jungeren-astrazeneca-geimpften-
        anderen-impfstoff-fur-zweite-dosis-UZK4LE65IVAVBK2XFZNHXKLLHE.html.

[102]   World Health Organization: https://www.who.int/influenza_vaccines_plan/resources/
        Session4_VEfficacy_VEffectiveness.PDF.

[103]   S. Black: 'Moderna reports 94.5 % efficacy for COVID-19 vaccine', Article, Science-
        board.net, 2020.

[104]   F. P. Polack, S. J. Thomas, N. Kitchin et al.: 'Safety and Efficacy of the BNT162b2
        mRNA COVID-19 Vaccine' in The New England Journal of Medicine, Volume 383,
        pp. 2603–2615,December 2020.

[105]   AstraZeneca: https://www.astrazeneca.com/content/astraz/media-centre/press-
        releases/2020/azd1222hlr.html.

[106]   Johns Hopkins University: https://coronavirus.jhu.edu/vaccines/vaccines-faq.

[107]   M. Levine-Tiefenbrun, I. Yelin, R. Katz et al.: 'Decreased SARS-CoV-2 viral load
        following vaccination' in medRxiv – PreprintServer for Health Sciences, 21251283,
        February 2021.

[108]   M. Voysey, S. A. Costa Clemens, S. A. Madhi et al.: 'Single dose administration, and
        the influence of the timing of the booster dose on immunogenicity and efficacy of

ChAdOx1 nCoV-19 (AZD1222) vaccine' in the Lancet Preprint, Volume 397, Issue 10277, pp. 881–891, March 2021.

[109] Nature.com: https://www.nature.com/articles/d41586-020-03334-w.

[110] World Health Organization: https://www.who.int/publications/m/item/draft-landscape-of-COVID-19-candidate-vaccines.

[111] I. Jones, P. Roy: 'Sputnik V COVID-19 vaccine candidate appears safe and effective' in The Lancet, Volume 20, Issue 397, pp. 642–643, February 2021.

[112] D. Logunov, I. V. Dolzhikova et al.: 'Safety and efficacy of an rAd26 and rAd5 vector-based heterologous prime-boost COVID-19 vaccine: an interim analysis of a randomised controlled phase 3 trial in Russia' in The Lancet, February 2021.

[113] J. Sadoff, M. Le Gars et al.: 'Interim Results of a Phase 1–2a Trial of Ad26.COV2.S COVID-19 Vaccine' in The New England Journal of Medicine, January 2021.

[114] Johnson and Johnson: https://www.jnj.com/johnson-johnson-announces-single-shot-janssen-COVID-19-vaccine-candidate-met-primary-endpoints-in-interim-analysis-of-its-phase-3-ensemble-trial.

[115] Bundesgesundheitsministerium: https://www.bundesgesundheitsministerium.de/coronavirus/faq-COVID-19-impfung.html#c20329.

[116] Robert-Koch Institute: https://www.rki.de/DE/Content/Infekt/Impfen/ImpfungenAZ/COVID-19/Impfstrategie_Covid19_Ueberblick.pdf?__blob=publicationFile.

[117] Deutsches Ärzteblatt: https://www.aerzteblatt.de/nachrichten/120637/Astrazeneca-Impfstoff-in-Deutschland-nur-fuer-Menschen-unter-65-Jahren.

[118] World Health Organization: https://www.who.int/news-room/feature-stories/detail/the-oxford-astrazeneca-COVID-19-vaccine-what-you-need-to-know.

[119] G. Waschinski et al.: 'Corona-Impfstoff: Diskussion um Wirksamkeit von Astra-Zeneca-Vakzin bei Senioren' in Handelsblatt, January 2021.

[120] Ourworldindata.org: https://ourworldindata.org/covid-vaccinations.

[121] T. Nierendorf: 'Reiche Länder wollen rücksichtslos kaufen' in Frankfurter Allgemeine Zeitung, December 2020.

[122] L. Yurkovetskiy, K. E. Pascal, C. Tomkins-Tinch: 'SARS-CoV-2 Spike protein variant D614G increases infectivity and retains sensitivity to antibodies that target the receptor binding domain' in bioRxiv – PreprintServer for Biology, 187757, July 2020.

[123] Robert-Koch Institute: https://www.rki.de/DE/Content/InfAZ/N/Neuartiges_Coronavirus/Virologische_Basisdaten.html.

[124] Helmholtz-Gemeinschaft Deutscher Forschungszentren : https://www.helmholtz.de/glossar/glossar-detail/spike-protein/.

[125] European Centre for Disease Prevention and Control: 'Rapid increase of a SARS-CoV-2 variant with multiple spike protein mutations observed in the United Kingdom ', December 2020.

[126] Y. J. Hou, S. Chiba et al.: 'SARS-CoV-2 D614G variant exhibits efficient replication ex vivo and transmission in vivo' in Science, Volume 370, Issue 6523, December 2020.

[127] World Health Organization: https://www.who.int/csr/don/31-december-2020-sars-cov2-variants/en/.

[128] N. G. Davies, R. C. Barnard, C. I. Jarvis et al.: 'Estimated transmissibility and severity of novel SARS-CoV-2 Variant of Concern 202012/01 in England' in medRxiv – PreprintServer for Health Sciences, 20248822, December 2020.

[129] A. Muik, A. Wallisch, B. Sänger et al.: 'Neutralization of SARS-CoV-2 lineage B.1.1.7 pseudovirus by BNT162b2 vaccine-elicited human sera' in bioRxiv – PreprintServer for Biology, 426984, January 2021.

[130] N. G. Davies et al.: 'Increased mortality in community-tested cases of SARS-CoV-2 lineage B.1.1.7' in medRxiv – PreprintServer for Health Sciences, 21250959, March 2021.

[131] Centers for Disease Control and Prevention: https://www.cdc.gov/coronavirus/2019-ncov/more/science-and-research/scientific-brief-emerging-variants.html.

[132] A. C. Williams, W. A. Burgers: 'SARS-CoV-2 evolution and vaccines: cause for concern?' in The Lancet Respiratory Medicine, January 2021.

[133] J. Wise: 'COVID-19: The E484K mutation and the risks it poses' in The British Journal of Medicine, February 2021.

[134] N. R. Faria1, I. Morales Claro, D. Candido et al.: 'Genomic characterisation of an emergent SARS-CoV-2 lineage in Manaus: preliminary findings' in Virological.org, January 2021.

[135] Deutsches Ärzteblatt: https://www.aerzteblatt.de/nachrichten/120353/Weitere-Mutationen-von-SARS-CoV-2-in-Deutschland-nachgewiesen.

[136] Association for Professionals in Infection Control and Epidemiology: https://apic.org/monthly_alerts/herd-immunity/.

[137] A. Fontanet, S. Cauchernez: 'COVID-19 herd immunity: where are we?' in Nature Public Health Emergency Collection, Volume 20, pp. 583–584, September 2020.

[138] Deutsches Ärzteblatt: https://www.aerzteblatt.de/nachrichten/118837/Fuer-Herdenimmunitaet-Coronaimpfrate-von-bis-zu-70-Prozent-noetig

[139] World Health Organization: https://www.who.int/news-room/q-a-detail/herd-immunity-lockdowns-and-COVID-19.

[140] Johns Hopkins Bloomberg School of Public Health: https://www.jhsph.edu/COVID-19/articles/achieving-herd-immunity-with-covid19.html.

[141] Nature.com: https://www.nature.com/articles/d41586-021-00121-z.

[142] K. Kupferschmidt: 'Vaccine 2.0: Moderna and other companies plan tweaks that would protect against new coronavirus mutations' in Science, January 2021.

[143] K. Wu, A. P. Werner, J. I. Moliva et al. 'mRNA-1273 vaccine induces neutralizing antibodies against spike mutants from global SARS-CoV-2 variants' in bioRxiv – PreprintServer for Biology, 427948, January 2021.

[144] Z. Wang, F. Schmidt, Y. Weisblum et al.: 'mRNA vaccine-elicited antibodies to SARS-CoV-2 and circulating variants' in bioRxiv – PreprintServer for Biology, 426911, January 2021.

[145] C. K. Wibmer, F. Ayres, T. Hermanus et al.: 'SARS-CoV-2 501Y.V2 escapes neutralization by South African COVID-19 donor plasma' in bioRxiv – PreprintServer for Biology, 18.427166, January 2021.

[146] Robert-Koch Institute: https://www.rki.de/DE/Content/Gesundheitsmonitoring/Gesundheitsberichterstattung/Glossar/gbe_glossar_catalog.html?cms_lv2=3686292.

[147] O. N. Bjornstad: 'Epidemics', Springer, pp. 18–33, Switzerland, 2018.

[148] B. S. Thomas, N. A. Marks: 'Estimating the Case Fatality Ratio for COVID-19 using a Time-Shifted Distribution Analysis' in medRxiv – PreprintServer for Health Sciences, 20216671, October 2020.

[149] Wikipedia: https://de.wikipedia.org/wiki/Virulenz.

[150] A. Stang, F. Standl, B. Kowall et al.: 'Excess mortality due to COVID-19 in Germany' in Journal of Infection, Volume 81, Issue 5, pp. 797–801, November 2020.

[151] Johns Hopkins University: https://coronavirus.jhu.edu/data/mortality.

[152] Ourworldindata.org: https://ourworldindata.org/coronavirus-data-explorer?zoomToSelection=true&minPopulationFilter=1000000&country=GBR~ITA~DEU~SWE&region=World&cfrMetric=true&interval=smoothed&hideControls=true&perCapita=true&smoothing=7&pickerMetric=location&pickerSort=asc

[153] EuroMOMO: https://www.euromomo.eu/graphs-and-maps.

[154] German Network for Evidence-based Medicine: https://www.ebm-netzwerk.de/en/publications/COVID-19.

[155] A. Stang, M. Stang, K. Jöckel: 'Estimated Use of Intensive Care Beds Due to COVID-19 in Germany Over Time' in Deutsches Ärzteblatt, Volume 117, pp. 329–135, May 2020.

[156] Statistisches Bundesamt: https://www.destatis.de/DE/Themen/Querschnitt/Corona/Downloads/dossier-COVID-19.pdf?__blob=publicationFile, last access on January 29th 2021, 11:20 am.

[157] Fraunhofer Institute for Industrial Mathematics: https://www.itwm.fraunhofer.de/en/departments/mf/latest-news/blog-streuspanne/corona-dark-figure.html.

[158] M. Schrappe, H. Francois-Kettner, M. Guhl, D. Hart, F. Knieps, P. Manow, H. Pfaff, K. Püschel, G. Glaeske: 'Die Pandemie durch SARS-CoV-2, COVID-19 - Datenbasis verbessern, Prävention gezielt weiterentwiceln, Bürgerrechte wahren', Thesenpapier, April 2020.

[159] NDR: https://www.ndr.de/nachrichten/hamburg/Forscher-und-Aerzte-kritisieren-die-Corona-Politik,pueschel310.html.

[160] C. Staerk, T. Wistuba, A. Mayr: 'Estimating effective infection fatality rates during the course of the COVID-19 pandemic in Germany' in arXiv, 2011.02420, November 2020.

[161] M. Catalá, D. Pino, M. Marchena et al.: 'Robust estimation of diagnostic rate and real incidence of COVID-19 for European policymakers' in PLOS ONE, Januar 2021.

[162] H. Nishiura, T. Kobayashi, Y. Yang, K. Hayashi et al.: 'The Rate of Underascertainment of Novel Coronavirus (2019-nCoV) Infection: Estimation Using Japanese Passengers Data on Evacuation Flights' in Journal of Clinical Medicine, Volume 9, Issue 2, p. 419, February 2020.

[163] A. Chiolero: 'Ranking lethality of COVID-19 and other epidemic diseases' in The British Journal of Medicine, Volume 369, April 2020.

[164] W. J. Guan et al.: 'Cardiovascular comorbidity and its impact on patients with COVID-19' in European Respiratory Journal, Volume 55, Issue 6, June 2020.

[165] B. Wang et al.: 'Does comorbidity increase the risk of patients with COVID-19: evidence from meta-analysis' in Aging, Volume 12, pp. 6049–6057, April 2020.

[166] A. B. Docherty et al.: 'Features of 16,749 hospitalised UK patients with COVID-19 using the ISARIC WHO Clinical Characterisation Protocol' in medRxiv – PreprintServer for Health Sciences, 20076042, April 2020.

[167] N. Stefan et al.: 'Obesity and impaired metabolic health in patients with COVID-19' in Nature Reviews Endocrinology, Volume 16, pp. 341–342, April 2020.

[168] L. Zhu et al.: 'Association of Blood Glucose Control and Outcomes in Patients with COVID-19 and Pre-existing Type 2 Diabetes' in Cell Press, Volume 31, Issue 6, June 2020, pp. 1068–1077.

[169] Q. Zhao et al.: 'The impact of COPD and smoking history on the severity of COVID-19: A systemic review and meta-analysis' in Journal of Medical Virology, Volume 92, pp. 1915–1921, October 2020.

[170] C. Vetter: 'COVID-19 bei Patienten mit Asthma oder COPD: Inhalative Steroide wirken bei COVID-19 nicht protektiv, sie schaden aber auch nicht' in Deutsches Ärzteblatt, Volume 118, Issue 1–2, January 2021.

[171] J. Leung et al.: 'COVID-19 and COPD' in European Respiratory Journal, Volume 56, Issue 2, August 2020.

[172] Statista.de: https://www.statista.com/statistics/1110949/common-comorbidities-in-COVID-19-deceased-patients-in-italy/.

[173] C. Edler, A. S. Schröder, M. Aepfelbacher et al. : 'Dying with SARS-CoV-2 infection-an autopsy study of the first consecutive 80 cases in Hamburg, Germany' in International Journal of Legal Medicine, Volume 134, Issue 4, July 2020.

[174] K. Püschel, M. Aepfelbacher: 'Umgang mit Corona-Toten: Obduktionen sind keinesfalls obsolet' in Deutsches Ärzteblatt, Volume 117, Issue 20, May 2020.

[175] Spiegel: https://www.berliner-kurier.de/panorama/kieler-pathologe-die-meisten-sterben-an-und-nicht-mit-corona-li.138122.

[176] Doc Check Flexikon: https://flexikon.doccheck.com/de/Infection_fatality_rate.

[177] A. T. Levin et al.: 'Assessing the age specifcity of infection fatality rates for COVID-19: systematic review, meta-analysis, and public policy implications' in European Journal of Epidemiology, Volume 35, pp. 1123–1138, December 2020.

[178] W. Yang et al.: 'Estimating the infection-fatality risk of SARS-CoV-2 in New York City during the spring 2020 pandemic wave: a model-based analysis' in The Lancet, Volume 21, Issue 2, pp. 203–212, October 2020.

[179] K. Kolenda: 'Letalität bei COVID-19 fünfmal höher als bei saisonaler Grippe' in Telepolis, December 2020.

[180] Y. Xie, B. Bowe, G. Maddukuri, Z. Al-Aly: 'Comparative evaluation of clinical manifestations and risk of death in patients admitted to hospital with COVID-19 and seasonal influenza: cohort study' in The British Journal of Medicine, Volume 371, December 2020.

[181] Robert-Koch Institute: 'Epidemiologisches Bulletin', October 2020.

[182] L. Piroth, J. Cottenet, A. Mariet, P. Bonniaud, M. Blot, P. Tubert-Bitter: 'Comparison of the characteristics, morbidity, and mortality of COVID-19 and seasonal influenza: a nationwide, population-based retrospective cohort study' in The Lancet Respiratory Medicine, December 2020.

[183] G. M. Fröhlich, M. De Kraker, M. Abbas et al.: 'COVID-19: more than 'a little flu'? Insights from the Swiss hospital-based surveillance of Influenza and COVID-19' in medRxiv – PreprintServer for Health Sciences, 20233080, November 2020.

[184] N. Siegmund-Schulze: 'Akutes respiratorisches Syndrom bei COVID-19: Autopsien zeigen einige pathophysiologische Unterschiede zu Influenza auf' in Deutsches Ärzteblatt, Volume 117, Issue 38, September 2020.

[185] Ourworldindata.org: https://ourworldindata.org/excess-mortality-covid.

[186] F. zur Nieden, B. Sommer, S. Lüken: 'Sonderauswertung der Sterbefallzahlen 2020', Statistisches Bundesamt, Wiesbaden, August 2020.

[187] Statistisches Bundesamt: https://www.destatis.de/EN/Themes/Society-Environment/ Population/Deaths-Life-Expectancy/mortality.html.

[188] European Centre for Disease Prevention and Control: https://www.ecdc.europa.eu/en/ seasonal-influenza/season-2017-18.

[189] Deutsches Ärzteblatt: https://www.aerzteblatt.de/nachrichten/117790/SARS-CoV-2- Von-der-Leyens-Plan-gegen-die-zweite-Welle.

[190] C. Hohmann-Jeddi: 'Auf die zweite Welle gut vorbereitet' in Pharmazeutische Zeitung, May 2020.

[191] W. Geissel: 'Dexamethason: WHO feiert Erfolg bei schwerer COVID-19' in Ärztezeitung, June 2020.

[192] M. A. Matthey, B. T. Thompson: 'Dexamethasone in hospitalised patients with COVID- 19: addressing uncertainties' in The Lancet Respiratory Medicine, Volume 8, Issue 12, pp. 1170–1172, December 2020.

[193] P. Horby, W.E. Lim, J. Emberson et al.: 'Effect of Dexamethasone in Hospitalized Patients with COVID-19 – Preliminary Report' in New England Journal of Medicine, Volume 384, pp. 693–704, July 2020.

[194] Statistisches Bundesamt: https://www.destatis.de/EN/Themes/Cross-Section/Corona/ Society/population_death.html.

[195] O. Diekmann, J. A. P. Heesterbeek, M. G. Roberts: 'The construction of next-generation matrices for compartmental epidemic models', Journal of the Royal Society Interface, pp. 873–885, June 2010.

[196] D. Brockmann, D. Helbing: 'The Hidden Geometry of Complex, Network-Driven Contagion Phenomena' in Science, Volume 342, Issue 6164, pp. 1337–1342, 2013.

[197] F. Iannelli, A. Koher, D. Brockmann, et al.: 'Effective distances for epidemics spreading on complex networks' in Physical Review E 95, 2017.

[198] W. O. Kermack, A. G. McKendrick: 'A contribution to the mathematical theory of epidemics' in Proceedings of The Royal Society A, Volume 115, pp. 700–721, 1927.

[199] S. Bansal, B. T. Grenfell, L. A. Meyers: 'When individual behaviour matters: homoge- neous and network models in epidemiology' in Journal of The Royal Society Interface, Volume 4, Issue 16, pp. 879–891, October 2007.

[200] M. Martcheva: 'An Introduction to Mathematical Epidemiology', Springer, New York, 2015.

[201] M. Li, X. Liu: 'An SIR Epidemic Model with Time Delay and General Nonlinear Incidence Rate' in Hindawi Publishing Corporation, December 2013.

[202] O. Arino, M. L. Hbid, E. A. Dads: 'Delay Differential Equations and Applications' in NATO Science Series II. Mathematics, Physics and Chemistry, Volume 205, Springer, September 2002.

[203] Wikipedia: https://en.wikipedia.org/wiki/Survival_analysis.

[204] M. Kocanczyk, F. Grabowski, T. Lipniacki: 'Dynamics of COVID-19 pandemic at constant and time-dependent contact rates' in Mathematical Modelling of Natural Phenomena, Voume 15, Article Number 28, 2020011, April 2020.

[205] W. Nagel: 'Vorlesungsskript: Wahrscheinlichkeitstheorie für Physiker', Jena, Septem- ber 2008.

[206] Wikipedia: https://en.wikipedia.org/wiki/Erlang_distribution.

[207] https://de.wikipedia.org/wiki/Erlang-Verteilung.

[208] D. Devianto, L Oktasari, Maiyastri: 'Some Properties of Hypoexponential Distribution with Stabilizer Constant' in Applied Mathematical Sciences, Volume 9, p. 7064, Indonesia, 2015.

[209] Wikipedia: https://fr.wikipedia.org/wiki/Loi_hypo-exponentielle.

[210] Wikipedia: https://en.wikipedia.org/wiki/Hyperexponential_distribution.

[211] Wikipedia: https://en.wikipedia.org/wiki/Hyperexponential_distribution.

[212] Actuarialmodelingtopics: https://actuarialmodelingtopics.wordpress.com/2016/08/01/the-hyperexponential-and-hypoexponential-distributions/.

[213] Infektionsschutz.de: https://www.infektionsschutz.de/coronavirus/fragen-und-antworten/quarantaene-und-isolierung.html#faq3788.

[214] Bundesgesundheitsministerium: https://www.bundesgesundheitsministerium.de/coronavirus-infos-reisende/faq-tests-einreisende.html#c20216.

[215] Robert-Koch Institute: https://www.rki.de/DE/Content/InfAZ/N/Neuartiges_Coronavirus/Kontaktperson/Management.html; jsessionid=424555F607E53EB7850B735EE92F3739.internet071?nn=13490888#doc13516162bodyText8.

[216] Robert-Koch Institute: https://www.rki.de/SharedDocs/FAQ/NCOV2019/FAQ_Liste_Kontaktpersonenmanagement.html.

[217] European Centre for Disease Prevention and Control: 'Contact tracing: public health management of persons, including healthcare workers, who have had contact with COVID-19 cases in the European Union – third update', November 2020.

[218] E. M. Anderson, E- C. Goodwin, A. Verma, C. P. Arevalo: 'Seasonal human coronavirus antibodies are boosted upon SARS-CoV-2 infection but not associated with protection' in medRxiv – PreprintServer for Health Sciences, 20227215, November 2020.

[219] Robert-Koch Institute, Germany: https://www.rki.de/DE/Content/InfAZ/N/Neuartiges_Coronavirus/Falldefinition.pdf?__blob=publicationFile.

[220] M. Ehrhardt, J. Gašper and S. Kilianová: 'SIR-based Mathematical Modeling of Infectious Diseases with Vaccination and Waning Immunity', Journal of Computational Science, Volume 37, 101027, October 2019.

[221] S. F. Lumley, D. O'Donell, N. E. Stoesser et al.: 'Antibody Status and Incidence of SARS-CoV-2 Infection in Health Care Workers' in The New England Journal of Medicine, pp. 533–540, December 2020.

[222] M. V. Barbarossa, G. Röst: 'Mathematical models for vaccination, waning immunity and immune system boosting: a general framework' in World Scientific, BIOMAT 2014, pp. 185–2015, January 2015.

[223] M. P. Dafilis, F. Frascoli, J. G. Wood, J. M. McCaw: 'The influence of increasing life expectancy on the dynamics of SIRS systems with immune boosting' in Anziam Journal, Volume 54, Issues 1–2, pp. 50–63, April 2013.

[224] Robert-Koch Institute: https://www.rki.de/SharedDocs/FAQ/COVID-Impfen, last access: February 3rd 2021, 10:30 am.

[225] C. A. Siegrist: 'Vaccine immunology' in Plotkin SA, Orenstein WA and Offit PA (eds.) Vaccines, Elsevier Inc, Philadelphia, 2008.

[226] R. P. Curiel, H. G. Ramírez: 'Vaccination strategies against COVID-19 and the diffusion of anti-vaccination views' in Nature Scientific Reports, Volume 11, Article Number 6626, March 2021.

[227] E. Estrada: 'The Structure of Complex Networks: Theory and Applications' in Oxford
       University Press, 2011.
[228] K. Shah, D. Saxena, D. Macalankar: 'Secondary Attack Rate of COVID-19 in house-
       hold contacts: Systematic review' in Oxford University Press, July 2020.
[229] Z. J. Madwell, Y. Yang, I. M. Longini: 'Household Transmission of SARS-CoV-2
       A Systematic Review and Meta-analysis' in JAMA Network, Volume 3, Issue 12,
       December 2020.
[230] Centers for Disease Control and Prevention: https://www.cdc.gov/flu/vaccines-work/
       effectivenessqa.htm.
[231] Centers for Disease Control and Prevention: https://www.cdc.gov/csels/dsepd/ss1978/
       lesson3/section2.html.
[232] A. Orenstein et al.: 'Field evaluation of vaccine efficacy' in Bulletin of the World
       Health Organization, Volume 63, Issue 6, pp. 1055–1068, 1985.
[233] Redaktionsnetzwerk Deutschland: https://www.rnd.de/gesundheit/stiko-expertin-
       zu-denken-nachstes-weihnachten-sei-alles-wieder-wie-fruher-halte-ich-fur-zu-
       optimistisch-TC5B3VAGIFDSZFH2KWYZZK5VAE.html.
[234] K. M. Bubar et al.: 'Model-informed COVID-19 vaccine prioritization strategies by
       age and serostatus', Boulder, December 2020.
[235] Max Planck Institute: https://www.mpic.de/4747361/risk-calculator.
[236] Y. Zeng et al.: 'Forecasting of COVID-19: spread with dynamic transmission rate' in
       Journal of Safety Science and Resilience, Tsinghua University in Beijing in China,
       pp. 91–96, December 2020.
[237] C. J. Carlson et al.: 'Misconceptions about weather and seasonality must not misguide
       COVID-19 response' in Nature Communications, Volume 11, Article Number 4312,
       August 2020.
[238] European Centre for Disease Prevention and Control: https://www.ecdc.europa.eu/en/
       COVID-19/latest-evidence/coronaviruses.
[239] T. To, K. Zhang, B. Maguire et al.: 'Correlation of ambient temperature and COVID-19
       incidence in Canada' in Science of the Total Environment, January 2021.
[240] M. Moriyama, W. J. Hugentobler, A. Iwasaki: 'Seasonality of Respiratory Viral Infec-
       tions', Annual Review of Respiratory Infections, Volume 7, pp. 83–101, March 2020.
[241] B. Tang et al.: 'The effectiveness of quarantine and isolation determine the trend of the
       COVID-19 epidemic in the final phase of the current outbreak in China' in International
       Journal of Infectious Diseases, Volume 96, pp. 288–293, June 2020.
[242] Wikipedia: https://en.wikipedia.org/wiki/Delay_differentialequation.
[243] T. Erneux: 'Applied Delay Differential Equations' in Surveys and Tutorials in the
       Applied Mathematical Sciences, Volume 3, Springer, 2009.
[244] P. Yan, S. Liu: 'SEIR Epidemic Model with Time Delay', Anziam Journal, Volume 48,
       pp. 121–123, 2006.
[245] A. Sommer: 'Numerical Methods for Parameter Estimation in Dynamical Systems
       with Noise', Universität Heidelberg.
[246] Mathworks: https://blogs.mathworks.com/cleve/2014/05/26/ordinary-differential-
       equation-solvers-ode23-and-ode45/#a495083d-5d4a-4fac-afbc-768444430751.
[247] Mathworks: https://de.mathworks.com/help/matlab/math/choose-an-ode-solver.
       html.

[248] M. Kimathi, S. Mwalili, V. Ojiambo Gathungu: 'Age-Structured Model for COVID-19: Effectiveness of Social Distancing and Contact Reduction in Kenya' in Infectious Disease Modelling, Volume 6, Issue 13, November 2020.

[249] Z. Zhao et al.: 'A five-compartment model of age-specific transmissibility of SARS-CoV-2' in Infectious Diseases of Poverty, Volume 9, Article Number 117, August 2020.

[250] Healthknowledge.orf.uk: https://Healthknowledge.org.uk/public-health-textbook/research-methods/1a-epidemiology/epidemic-theory.

[251] A. B. Gumel et al.: 'Modelling strategies for controlling SARS outbreaks' in Proceedings of the Royal Society, Biological Sciences Series B, Volume 271, Issue 1554, rspb.2004.2800, November 2004.

[252] Robert-Koch Institute: 'Erläuterung der Schätzung der zeitlich variierenden Reproduktionszahl R', May 2020.

[253] Statista.de: https://de.statista.com/statistik/daten/studie/1117478/umfrage/reproduktionszahl-des-coronavirus-COVID-19-in-deutschland/.

[254] A. Perasso: 'An introduction to the Basic Reproduction number in mathematical epidemiology' in Mathe-matical Modelling and Numerical Analysis: Proceedings and Survey, Volume 62, pp. 123–138.

[255] M. V. Barbarossa, N. Bogya, A. Dénes G. Röst, H. V. Varma, Z. Vizi: 'Fleeing lockdown and its impact on the size of epidemic outbreaks in the source and target regions – a COVID-19 lesson' in Research Square, September 2020.

[256] G. O. Fosu, E. Akweittey: 'Next-Generation Matrices and Basic Reproduction Numbers for All Phases of the Coronavirus Disease' in Open Journal of Mathematical Sciences, pp. 261–272, May 2020.

[257] N. Chintis, J.M. Cushing, J.M. Hyman: 'Determining Important Parameters in the Spread of Malaria Through the Sensitivity Analysis of a Mathematical Model', Bulletin of Mathematical Biology, Volume 70, Issue 5, pp. 1272–1296, August 2008.

[258] Diehl, M.: 'Lecture Notes on Numerical Optimization' (Preliminary Draft), pp. 37–72, Freiburg, March 2016.

[259] A. Melegaro et al.: 'What types of contacts are important for the spread of infections? Using contact survey data to explore European mixing patterns' in Epidemics, Volume 3, Issues 3–4, pp. 143–151, September 2020.

[260] Statistisches Bundesamt: https://www.destatis.de/DE/Themen/Gesellschaft-Umwelt/Bevoelkerung/Bevoelkerungsstand/_inhalt.html.

[261] Statistikmyndigheten SCB: https://www.scb.se/hitta-statistik/statistik-efter-amne/befolkning/befolkningens-sammansattning/befolkningsstatistik/.

[262] Statistisches Bundesamt: https://www.destatis.de/DE/Themen/Gesellschaft-Umwelt/Bevoelkerung/Sterbefaelle-Lebenserwartung/_inhalt.html.

[263] Findikator: https://findikaattori.fi/sv/46.

[264] Statista.de: https://de.statista.com/statistik/daten/studie/1190592/umfrage/coronainfektionen-und-hospitalisierte-faelle-in-deutschland-nach-meldewoche/#professional.

[265] Statista.de: https://www.statista.com/statistics/1105753/cumulative-coronavirus-deaths-in-sweden/.

[266] European Centre for Disease Prevention and Control: https://www.ecdc.europa.eu/en/publications-data/download-data-hospital-and-icu-admission-rates-and-current-occupancy-COVID-19.

[267] Statista.de: https://de.statista.com/statistik/daten/studie/1013307/umfrage/sterbefaelle-in-deutschland-nach-alter/.

[268] Statista.de: https://de.statista.com/statistik/daten/studie/1104173/umfrage/todesfaelle-aufgrund-des-coronavirus-in-deutschland-nach-geschlecht/.

[269] Robert-Koch Institute: https://www.rki.de/DE/Content/InfAZ/N/Neuartiges_Coronavirus/Daten/Impfquoten-Tab.html;jsessionid=7E37C0BBB9A1EDA08191A7E1AC221D1B.internet051.

[270] Statistisches Bundesamt: https://www.destatis.de/DE/Presse/Pressemitteilungen/2020/12/PD20_N082_122.html.

[271] Intensivregister.de: https://www.intensivregister.de/#/aktuelle-lage/zeitreihen.

[272] Worldometers.info: https://www.worldometers.info/coronavirus/country/germany/.

[273] Statista.de: https://www.statista.com/statistics/1102203/cumulative-coronavirus-cases-in-sweden/.

[274] Statista.de: https://www.statista.com/statistics/1105753/cumulative-coronavirus-deaths-in-sweden/.

[275] Folkshälsomyndigheten: https://www.folkhalsomyndigheten.se/smittskydd-beredskap/utbrott/aktuella-utbrott/COVID-19/statistik-och-analyser/analys-och-prognoser/.

[276] Robert-Koch Institute: https://www.rki.de/DE/Content/InfAZ/N/Neuartiges_Coronavirus/Daten/Altersverteilung.html.

[277] R. E. Mickens: 'Applications of Nonstandard Finite Difference Schemes' in World Scientific, March 2000.

[278] U. Irkhin, M. Lapinska-Chrzczonowicz: 'Exact difference schemes for time-dependent problems' in Computational Methods in Applied Mathematics, Volume 5, Issue 4, pp. 442–448, 2005.

[279] A. Suryanto: 'A Conservative Nonstandard Finite Difference Scheme for SIR Epidemic Model', Conference Paper, International Conferences and Workshops on Basic and Applied Sciences 2011, January 2011.

[280] M. Ehrhardt, R. E. Mickens: 'A Nonstandard Finite Difference Scheme for Solving a Zika Virus Model', unpublished manuscript.

[281] North Carolina State University: https://www4.stat.ncsu.edu/~gross/BIO56020webpage/slides/Jan102013.pdf.

[282] V. Marmará: 'Prediction of infectious disease outbreaks based in limited information', p. 6, University of Stirling, September 2016.

[283] W.C. Roda, M. B. Varughese, D. Han, M. Y. Li: 'Why is it difficult to accurately predict the COVID-19 epidemic?' in Infectious Disease Modelling, KeAi, Volume 5, pp. 271–281, March 2020.

[284] F. Liu, X. Li, G. Zhu: 'Using the contact network model and Metropolis-Hastings sampling to reconstruct the COVID-19 spread on the "Diamond Princess"' in Science Bulletin, Volume 65, Issue 15, pp. 1297–1305, August 2020.

[285] R. Mbuvha, T. Marwala: 'Bayesian Inference of COVID-19 Spreading Rates in South Africa' in medRxiv – PreprintServer for Health Sciences, 20083873, April 2020.

[286] Z. M. Zingoni et al.: 'A Critical Review of Multistate Bayesian Models Modelling Approaches in Monitoring Disease Progression: Use of Kolmogorov-Chapman Forward Equations with in WinBUGS' in ARC Journal of Public Health and Community Medicine, Volume 4, Issue 4, pp. 9–22, 2019.

[287] F. Brauer, P. van den Driessche, J. Wu: 'Mathematical Epidemiology' in Mathematical Biosciences Subseries, Springer, 2008.

[288] C. Lefèvre, M. Simon: 'SIR-Type Epidemic Models as Block-Structured Markov Processes' in Methodology and Computing in Applied Probability, Volume 22, pp. 433–453, April 2019.

[289] S. Tappe: 'Vorlesungsskript Markov-Ketten', Freiburg, 2018.

[290] Wikipedia: https://en.wikipedia.org/wiki/Failure_rate.

[291] C. Hougaard: 'Multi-state Models: A Review' in Lifetime Data Analysis, Volume 5, pp. 239–264, 1999.

[292] B. Singer, S. Spilerman: 'The Representation of Social Processes by Markov Models' in American Journal of Sociology, Volume 82, pp. 1-54, July 1976.

[293] Towardsdatascience.com: https://towardsdatascience.com/bayesian-inference-problem-mcmc-and-variational-inference-25a8aa9bce29.

[294] L. Fahrmeir, T. Kneib, S. Lang: 'Regression – Modelle, Methoden, Anwendungen', Springer, Volume 2, pp. 467–483, 2009.

[295] F. Weidemann: 'Bayesian Inference for Infectious Disease Transmission Models Based on Ordinary Differential Equations', Dissertation, Ludwigs-Maximilians-Universität, München, 2015.

[296] J. B. Johnson, K. S. Omland: 'Model selection in ecology and evolution' in Trends in Ecology & Evolution, Volume 19, pp. 101–108, February 2004.

[297] S. M. Lynch: 'Introduction to Applied Bayesian Statistics and Estimation for Social Scientists', Springer, p. 146, 2007.

[298] W.C. Roda: 'Bayesian inference for dynamical systems' in Infectious Disease Modelling, KeAi, Volume 5, pp. 221–232 , January 2020.

[299] H. Haario, E. Saksman, J. Tamminen: 'An adaptive Metropolis algorithm' in Project Euclid, Volume 7, Issue 2, 2001.

[300] T. F. Coleman, Y. Li: 'An Interior, Trust Region Approach for Nonlinear Minimization Subject to Bounds' in SIAM Journal of Optimization, Volume 6, pp. 418–445, 1996.

[301] G. Gonzalez-Parra, D. Martínez-Rodríguez, R. J. Villanueva-Micó: 'Impact of a New SARS-CoV-2 Variant on the Population: A Mathematical Modeling Approach' in Mathematical and Computational Applications, Volume 26, Issue 25, 26020025 , March 2021.

[302] T. House: 'Modelling Epidemics on Networks' in Contemporary Physics, Volume 53, Issue 3, November 2011.

[303] S. Lin, Y. Qiao, J. Huang, N. Yan: 'Research on the Influence of Effective Distance Between Cities on the Cross-regional Transmission of COVID-19' in medRxiv – PreprintServer for Health Sciences, 20044958, March 2020.

All online sources for which no last access date and time is specified were accessed for the last time on March 27th 2021 between 10:30 and 11:00 am.

Printed in the United States
by Baker & Taylor Publisher Services